Nancy P. Neas

DR. NANCY P. NEAS
DEPARTMENT OF BIOLOGY
COLGATE UNIVERSITY
HAMILTON, NY 13346

D0217409

Lipids in Plants and Microbes

TITLES OF RELATED INTEREST

The use of radioactive isotopes in the life sciences
J. M. Chapman & G. Ayrey

Class experiments in plant physiology
H. Meidner

Cover photograph: An electron micrograph of a
mixture of diacylgalactosylglycerol and
diacyldigalactosylglycerol, prepared by Mr A. Foley
(Department of Biochemistry, University College, Cardiff)
and photographed by Dr A. P. R. Brain (Department of Biochemistry,
Chelsea College, London).

Lipids in Plants and Microbes

JOHN L. HARWOOD
and
NICHOLAS J. RUSSELL

Department of Biochemistry,
University College of Wales, Cardiff

London
GEORGE ALLEN & UNWIN
Boston   Sydney

© J. L. Harwood and N. J. Russell, 1984
This book is copyright under the Berne Convention. No reproduction
without permission. All rights reserved.

George Allen & Unwin (Publishers) Ltd,
40 Museum Street, London WC1 1LU, UK

George Allen & Unwin (Publishers) Ltd,
Park Lane, Hemel Hempstead, Herts HP2 4TE, UK

Allen & Unwin Inc.,
9 Winchester Terrace, Winchester, Mass. 01890, USA

George Allen & Unwin Australia Pty Ltd,
8 Napier Street, North Sydney, NSW 2060, Australia

First published in 1984

British Library Cataloguing in Publication Data

Harwood, John L
    Lipids in plants and microbes.
1. Microbial lipids   2. Plant lipids
I. Title   II. Russell, Nicholas J.
576′.119293   QR92.L5
ISBN 0-04-574021-6
ISBN 0-04-574022-4 Pbk

Library of Congress Cataloging in Publication Data

Harwood, John L.
    Lipids in plants and microbes.
Bibliography: p.
Includes index.
1. Plant lipids. 2. Microbial lipids. I. Russell,
Nicholas J. II. Title.
QK898.L56H37   1984   581.19′247   84-9156
ISBN 0-04-574021-6 (alk. paper)
ISBN 0-04-574022-4 (pbk. : alk. paper)

Set in 9 on 11 point Times by Preface Ltd, Salisbury, Wilts
and printed in Great Britain by Butler & Tanner Ltd, Frome and London

Preface

This short text is designed to provide basic information about plant and microbial lipids not only for scientists working in the microbiological and plant fields, but for anyone wanting a concise introduction to this aspect of lipid biochemistry. We have long been aware that standard biochemistry books tend to concentrate (sometimes exclusively) on animal lipids, thus neglecting many of the important and special features of other organisms.

It is not our intention that the book should be comprehensive and we have not, for instance, provided complete lists of lipid compositions of all plants and bacterial species; a number of excellent specialist texts exist and many of these are listed for further reading. Instead we have sought to provide sufficient information for an advanced undergraduate or a research student to give them a 'feel' for the subject. By a combination of generalisation and the use of examples of special interest we hope the book will whet the appetite of the reader so that, by their own research, they are stimulated to discover and, perhaps, answer some of the fascinating questions concerning plant and microbial lipids. We trust that we shall succeed in these aims, even if that will mean more competition for research funds in our own fields!

J. L. HARWOOD
N. J. RUSSELL
November 1983

Acknowledgements

Our research careers have been devoted to a study of lipids: we have no regrets and are happy to acknowledge Professors J. N. Hawthorne, University of Nottingham, P. K. Stumpf, University of California, and E. F. Gale, University of Cambridge, who by their example and enthusiasm first set us on our respective courses and introduced us to the fascinating variety of different organisms. We are also grateful to all those colleagues and students who have helped us to develop these interests through research projects and other collaborative ventures, and thus indirectly helped to spawn this book. More specifically we should like to express our sincere thanks to Professor J. N. Hawthorne, University of Nottingham, and Professor M. Kates, University of Ottawa, who made helpful suggestions in the planning and early stages of writing the book; particular thanks go also to Drs R. M. C. Dawson and M. Goodfellow, who read the complete manuscript and made many constructive comments. Any errors that remain are the authors' responsibility entirely. We should also like to thank all those who supplied unpublished material, particularly Dr R. Anderson, University of Western Ontario, Dr P. B. Comita, University of Stanford, and Dr W. D. Grant, University of Leicester.

We wish to thank Mrs Barbara Power and Miss Jean Mitchell for efficiently converting our handwriting into legible typescript. We are very grateful to Mr Miles Jackson and Mr Geoffrey Palmer for their helpful advice and editorial assistance at George Allen & Unwin.

Finally, but by no means least, we should like to acknowledge the support of our wives and children; without their forebearance of lost evenings and weekends this book could not have been written, and we dedicate it to them.

J. L. H.
N. J. R.

Contents

1
Introduction

1A   General remarks

During the early growth of biochemistry relatively little attention was devoted to the study of lipids, apart from some specialised aspects like the lipid-soluble vitamins. In the past 20 years there has been a marked upsurge of interests in lipids, stimulated by research in areas such as photosynthesis and membrane structure and function. This research has encompassed the lipids of animals, plants and micro-organisms. However, despite the great economic importance of, for example, seed oils and the wide use of bacterial systems as experimental models, it is often assumed that lipids of plants and bacteria are essentially the same as those of animals. Alternatively, scant mention only is made of the differences.

In similar vein the impression may be given that 'everyman's bacterium', *Escherichia coli*, is the blueprint for all bacteria. The same might be said of spinach in the plant world! Nothing could be further from the truth and the diversity of lipid types and functions in the plant and bacterial kingdoms is both staggering and stimulating. Certain lipids are unique to a single species or type of plant or bacterium. Apart from their intrinsic interest as a rich source of research material for the lipid biochemist, they are of exceptional importance economically, especially those of plants, which provide enormous financial revenue from, for example, seed oils. With the advent of genetic engineering we can envisage increasing use being made of unicellular algae, bacteria and yeasts as providers of economically important lipid materials in areas such as the food and pharmaceutical industries.

One aim of this book is to provide an insight into the diversity of lipids in plants and bacteria. Comparisons are drawn with the lipids of animals, so as to complement more generally available information. Mention is made also of yeasts, algae and fungi. We do not intend to provide a compendium of the lipid compositions present in all bacteria and plants – that is beyond the scope of a concise text. Instead, a middle course is steered and an attempt is made to generalise whenever possible. If you are persuaded to read on, it will soon become clear that in this topic there is an exception to prove every rule! Particular mention is made of certain plant or bacterial species where these have special interest. For example, the extremely halophilic bacteria are used extensively as a biochemical system to study membrane phenomena, including energy production and transport. As such they are of fundamental interest and it is important, therefore, to deal with their lipids, which are unique.

1B Classification of organisms

One of the most basic divisions between all organisms is that between prokaryotes and eukaryotes. The prokaryotes include the bacteria and the cyanobacteria (or blue-green algae, as they were called formerly); all other organisms, including yeasts and fungi are eukaryotes – i.e. the term 'microbial' is not synonymous with 'prokaryotic'. The major distinction between prokaryotes and eukaryotes is the relative lack of intracellular organelles and absence of a nuclear membrane in prokaryotes (see Ch. 3). Recent studies on the sequences of 16s and other small RNA molecules, however, have revealed the existence of a third phylogenetic group (ur-kingdom) – the **archaebacteria**. These are made up of the extreme halophiles and some thermoacidophiles, thermoalkaliphiles and methanogens. This group of bacteria is believed to have originated from a quite distinct ancestor, and, interestingly from our point of view, its members contain quite unique lipids. Unless otherwise stated the term 'bacteria' is used in this book to refer to prokaryotes other than the archaebacteria and cyanobacteria. Bacteria are classified in 11 orders, which are further subdivided into genera and species. The broad outlines of this classification are shown in Table 1.1 for the benefit of those unfamiliar with bacterial types. This is not intended to be comprehensive, but rather to give a simple framework on which to place the subsequent information on bacterial lipids.

The two orders Pseudomonadales and Eubacteriales contain most of the bacterial species that have been well studied biochemically. The other nine orders consist of bacteria with some special attribute, and examples are given in Table 1.1. One of the most important from our point of view is the actinomycetes, which include the genera *Nocardia* and *Mycobacterium* that are characterised by their large content of distinctive waxy lipids. The genus *Corynebacterium*, previously classified in the Eubacteriales, is now included in the actinomycetes. This group is referred to subsequently as 'mycobacteria and related organisms'. This example illustrates the difficulty of much bacterial classification, which was based originally on factors such as morphology and growth characteristics. Newer biochemical information, including lipid composition (see Sec. 3A), has led to a certain amount of reclassification of organisms.

Another way of grouping bacteria is on the basis of the Gram staining reaction. Although there exists a continuous spectrum of reaction, essentially two groups can be identified, the so-called Gram-positives and Gram-negatives. This differential

Table 1.1 A simplified classification of bacteria showing some orders with representative genera.

A. Cyanobacteria
B. Archaebacteria
C. Bacteria (11 orders)
      (a) Pseudomonadales, e.g. photosynthetic bacteria, *Vibrio*, and *Pseudomonas*
      (b) Eubacteriales, e.g. *Escherichia, Salmonella, Micrococcus,* and *Bacillus*
      (c) nine 'specialised' orders, e.g. Actinomycetales (such as *Nocardia* and *Mycobacterium*), Myxobacteriales (gliding bacteria), Hyphomicrobiales (budding bacteria), Mycoplasmatales (wall-less bacteria), and Spirochaetales (flexible bacteria)

(a)

(b)

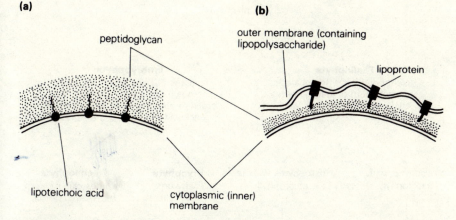

peptidoglycan

outer membrane (containing lipopolysaccharide)

lipoprotein

lipoteichoic acid

cytoplasmic (inner) membrane

Figure 1.1 A comparison of the morphology and lipid distribution in the cell walls of (a) Gram-positive and (b) Gram-negative bacteria.

staining reaction has a morphological basis in the structure of the cell wall. The most distinctive feature is the presence of a second, lipopolysaccharide-containing, membrane outside the cytoplasmic (inner) membrane in Gram-negatives but not Gram-positives (Fig. 1.1). There are also other qualitative, as well as quantitative, differences in the lipids of Gram-negatives and Gram-positives. Some bacterial orders (see Table 1.1) contain both Gram-positive and Gram-negative genera (e.g. Eubacteriales), whereas others contain only Gram-positive (e.g. Actinomycetales) or Gram-negative (e.g. Pseudomonadales) genera.

The cyanobacteria, although they are much larger than other prokaryotes, are not to be confused with eukaryotic algae. Cyanobacteria contain chlorophyll, not bacteriochlorophyll as in the green and purple photosynthetic bacteria. Eukaryotic algae were for many years divided into four groups based on their most conspicuous pigments – i.e. the blue-greens, greens, browns and reds. Modern classification still takes into account the occurrence of certain pigments but, in addition, emphasises features such as cytology, life histories, and the chemical nature of the cell wall. The number of divisions has, therefore, multiplied (depending on the author!) but six commonly used ones are shown in Figure 1.2 (seven divisions if one counts the blue-green algae, which are prokaryotic). The algae are all thallophyta but differ from yeasts, fungi and slime moulds which do not contain chlorophyll. Higher plants, mosses and ferns belong to the embryophyta, as shown in Figure 1.2.

Finally, it is appropriate at this point to mention a relationship between bacteria and the mitochondria of eukaryotes proposed by the endosymbiont hypothesis. This states that mitochondria arose during evolution by the engulfment of one primitive prokaryote by another, with subsequent differentiation of the endosymbiont into mitochondria. Chloroplasts might have arisen similarly from cyanobacteria. Some similarities in the lipid composition and metabolism of bacteria and (inner) mitochondrial membranes (or of chloroplast membranes and cyanobacteria) have been used as evidence for the endosymbiont theory, and these are pointed out where appropriate.

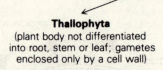

plants

Thallophyta
(plant body not differentiated
into root, stem or leaf; gametes
enclosed only by a cell wall)

Embryophyta
(zygotes develop into multicellular
embryo while within female sex
organ or within embryo sac; gametes
enclosed by a layer of sterile cells)

thallophytes with
chlorophyll

thallophytes without
chlorophyll

Bryophyta
(liverworts,
mosses)

Tracheophyta
(includes club mosses,
ferns, seed plants)

Chlorophyta
(green algae)

Myxomycota
(slime molds)

Euglenophyta
(euglenoids)

Eumycota
(includes algal fungi,
sac fungi, club fungi,
fungi imperfecti)

Chrysophyta
(yellow-green
algae, golden-
brown algae,
diatoms)

Pyrrhophyta
(dinoflagellates)

Phaeophyta
(brown algae)

Figure 1.2 A simplified classification of plants. Note that cyanophyta (blue-green algae) were formerly classified as plants, but are now considered to be bacteria (i.e. cyanobacteria); see Table 1.1.

Rhodophyta
(red algae)

1C Lipid nomenclature

To help readers, it is also appropriate to make a few comments about nomenclature and conventions regarding lipids. Trivial and abbreviated names are extensively used in text-books (and papers), because often these are more convenient than systematic names and they convey concisely a large amount of information. However, they may be incomprehensible and bewildering to a newcomer to the field. We have used such terms in this book only when they are widely used (and understood) or where there are technical reasons for doing so (e.g. the use of fatty acid abbreviations when the acids have only been partly identified).

A shorthand nomenclature is in common usage for fatty acids. They are written as two numbers separated by a colon. The number before the colon indicates the

Table 1.2 Examples of fatty acids and their abbreviations.

Trivial name	Symbol	Structure	Systematic name
palmitic acid	16 : 0	$CH_3(CH_2)_{14}COOH$	hexadecanoic acid
oleic acid	18 : 1 (9c)	$CH_3(CH_2)_7CH \overset{c}{=} CH(CH_2)_7COOH$	cis-9-octadecenoic acid
	15 : 0 iso-br	$CH_3CH(CH_3)(CH_2)_{11}COOH$	iso-14-methyltetradecanoic acid
lactobacillic acid	19 : 0 cyc	$CH_3(CH_2)_5CH-CH(CH_2)_9COOH$ $\underset{CH_2}{\diagup\diagdown}$	cis,cis-11,12-methyleneoctadecanoic acid
α-linolenic acid	18 : 3 (9c, 12c, 15c)	$CH_3CH_2CH \overset{c}{=} CHCH_2CH \overset{c}{=} CHCH_2CH \overset{c}{=} CH(CH_2)_7COOH$	all cis-9,12,15-octadecatrienoic acid

carbon chain length and the figure after corresponds to the number of double bonds. Additional figures in parenthesis show the position of double bonds and the letters *c, t* and *a*, show whether the bond is *cis*-olefinic, *trans*-olefinic or acetylenic. The position of the double bond is normally numbered from the carboxyl group (Δ numbering) but where it is more meaningful to number from the methyl end it is prefixed by ω. Branched-chain fatty acids are denoted by *br*, and cyclopropane fatty acids by *cyc*. These abbreviations are also shown in Table 1.2, which lists some important fatty acids. It must be stressed that an often used notation such as 18 : 1 merely means an octadecenoic acid and is *not* synonymous with a specific molecule (such as oleic acid) without further qualification.

Before we conclude this introductory chapter, two more items should be mentioned. Firstly, acyl lipids based on glycerol exhibit stereochemistry in much the same way as amino acids or monosaccharides. According to international convention, phosphoglycerides, for example, are named by *stereochemical numbering* with the letters *sn* being added to indicate this. Thus naturally occurring phosphatidylcholine becomes 1,2-diacyl-*sn*-glycero-3-phosphocholine. Secondly, we have tried to avoid any confusion between numbering total carbons and specific carbons in a compound in the following way: a number *before* C indicates a total number of carbon atoms but a number *after* C shows the specific carbon referred to. For example, oleic acid (Table 1.2) is an 18C acid with a double bond between C9 and C10 (numbering from the carboxyl end).

Lauric 12:0
Myristic 14:0
Palmitic 16:0
Stearic 18:0

2

Major lipid types in plants and micro-organisms

2A Fatty acids

Over 500 fatty acids have been found in plants and micro-organisms but only a few of them are important quantitatively. Because of their manner of synthesis (Sec. 4A.1), many fatty acids are even-chain saturated or unsaturated monocarboxylic acids. The majority are not found as free acids, but as constituents of more complex molecules such as acyl lipids and lipopolysaccharides. These will be dealt with after we have considered the various types of fatty acids.

In plants, seven types account for approximately 95% of the total fatty acids of leaf tissues or most commercial seed oils: these are saturated (lauric, myristic, palmitic, stearic) and unsaturated (oleic, linoleic and α-linolenic) acids (Table 2.1). The three unsaturated fatty acids all contain a *cis* double bond nine carbons from the carboxyl group (Δ9), and the polyunsaturated acids have their double bonds separated by a methylene group. The *cis* double bonds distort the acyl chain and contribute to membrane fluidity (Secs 6A.1 & 6A.2).

The most common substituted fatty acids in plants are monohydroxy derivatives – although polyhydroxy, keto, epoxy, dicarboxylic and even fluro acids are found. Usually, the monohydroxy acids are 2-derivatives (i.e. α-hydroxy acids) and they are found esterified in sphingolipids. In seed oils, the hydroxy fatty acids accumulate in triacylglycerol. The other derivatives appear to be important in surface layers (Sec. 2E.1).

Although hydroxy fatty acids are generally minor components of fungi, in certain mushrooms they may represent up to a quarter of the total fatty acids. In addition, several yeasts or yeast-like fungi produce extracellular hydroxy (and acetylated and long-chain) fatty acids.

Branched-chain acids such as cyclopropane (Fig. 2.1), cyclopentene and cyclopropene derivatives are found in some plants and fungi but their occurrence is rare when compared to bacteria. In lichens, however, the so-called lichen acids, which are cyclic aliphatic acids, are important perhaps because of their resistance to photo-oxidation.

Whereas the stem, roots and leaves of higher plants and most of their fruits contain only a few fatty acid types, the seeds of some plants accumulate unusual acids. Several of these (e.g. castor bean, rape) are important as sources of oils (see Sec. 6B 2).

Other photosynthetic organisms such as green and brown algae contain similar

α linolenic (plant-like) 18:3 ω3
γ linolenic (animal-like) 18:3 ω6

Table 2.1 Major fatty acids of plants and micro-organisms.

Trivial name	Symbol	Systematic name	Occurrence
lauric acid	12:0	dodecanoic acid	P
myristic acid	14:0	tetradecanoic acid	P, B, G, C, F, Br
β-hydroxymyristic acid	β-OH 14:0	3-hydroxytetradecanoic acid	B
methyltetradecanoic acid	15:0br	e.g. 13-methyltetradecanoic acid	B
palmitic acid	16:0	hexadecanoic acid	P, B, G, C, F, Br
methylhexadecanoic acid	17:0br	e.g. 14-methylhexadecanoic acid	B
—	17:0cyc	cis-9,10-methylene hexadecanoic acid	B
stearic acid	18:0	octadecanoic acid	P, B, G, F, Br
oleic acid	18:1(9c)	cis-9-octadecenoic acid	P, G, C, F, Br
vaccenic acid	18:1(11c)	cis-11-octadecenoic acid	B
linoleic acid	18:2(9c,12c)	cis,cis-9,12-octadecadienoic acid	P, G, C, F, Br
α-linolenic acid	18:3(9c,12c,15c)	all cis-9,12,15-octadecatrienoic acid	P, G, C, F, Br
lactobacillic acid	19:0cyc	cis-11,12-methylene octadecanoic acid	B
arachidonic acid	20:4(5c,8c,11c,14c)	all cis-5,8,11,14-eicosatetraenoic acid	C, G, F, Br
—	20:5(5c,8c,11c,14c,17c)	all cis-5,8,11,14,17-eicosapentaenoic acid	C, Br

Abbreviations: P, plants; B, bacteria; G, green algae; C, cyanobacteria; F, fungi; Br, brown algae.

acids to higher plants. However, marine algae often contain high amounts of very-long-chain polyunsaturated acids (e.g. arachidonic, 20:4, or eicosapentenoic, 20:5). Cyanobacteria are very diverse: some are 'animal-like' (containing γ-linolenic acid, 18:3 (6c,9c,12c)) while others are 'plant-like' (containing α-linolenic

general formula	type of fatty acid

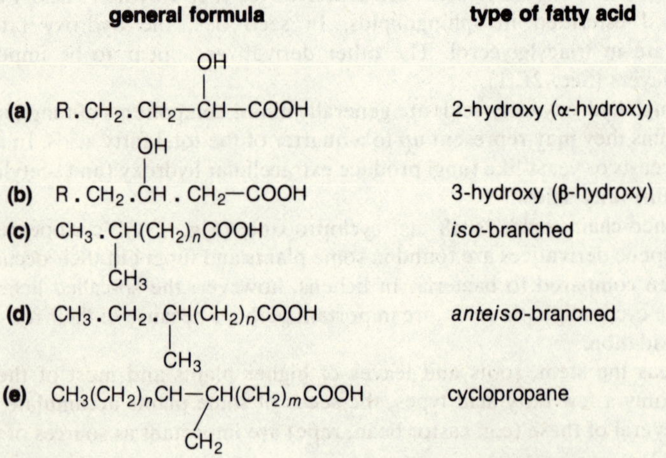

Figure 2.1 Important substituted fatty acids in plants and bacteria.

acid 18 : 3 $(9c,12c,15c)$). No cyanobacteria contain the unusual *trans*-Δ3-hexa- 16:1(n-12)
decenoic acid which is characteristic of the photosynthetic membranes of
eukaryotic algae and plants.

Fungi, like cyanobacteria, can also be divided on the basis of their fatty acid
composition. Although all fungi contain palmitic, stearic and unsaturated 18C fatty
acids, the lower fungi (phycomycetes) usually contain γ-linolenic acid while the
higher fungi (e.g. yeasts) contain α-linolenic acid. As usual there are exceptions and
a few fungi contain both! Arachidonic acid is a major component of a few fungal
species but very-long-chain acids are usually minor. Aquatic fungi often possess
distinct fatty acid patterns with as much as 75% of their total acids being
represented by one type.

Generally, the commonest bacterial fatty acids are 12–20C saturated and
monounsaturated types. A significant difference between bacteria and most other
organisms is that bacteria do not usually synthesise polyunsaturated fatty acids (a
notable exception being the gliding bacteria with large amounts of arachidonic
acid). Traces of polyunsaturated fatty acids may be present as contaminants derived
from growth medium constituents such as yeast extract or serum. Besides the
ubiquitous even-chain saturated and unsaturated fatty acids, bacteria characteristically contain odd-chain and branched fatty acids as well as 2- or 3-hydroxy and
cyclopropane derivatives (Table 2.1, Fig. 2.1), all of which are much less common
in higher organisms. Apart from their presence in membrane phospholipids and
glycolipids, bacterial fatty acids are found in lipopolysaccharide, cell wall lipoprotein and lipoteichoic acid (see Sec. 4D). The main differences between the fatty
acids of Gram-negative and Gram-positive bacteria are discussed in Section 3A.

The presence of *iso* or *anteiso* branched-chain acids (Fig. 2.1.) in membrane   branched
lipids has a similar effect in increasing fluidity as does the presence of *cis* double   fat fluid
bonds. Branched-chain acids are rarely unsaturated as well. Certain bacteria (e.g.
mycobacteria) contain exceptionally long-chain (up to 56C) branched-chain acids;
a number of these, like the phthioceranic acids (31–46C), have five to ten methyl
branches. Those mycobacteria that cause tuberculosis contain mycocerosic acids
which have a straight-chain portion of 18–20C and a multimethyl-branched portion
at the carboxyl end; the best characterised is 2,4,6,8-tetramethyloctacosanoic acid
from *Mycobacterium tuberculosis*. All these branched acids are found in trehalose
lipids in the waxy outer coats of mycobacteria, which also contain complex long-
chain 2-alkyl-branched 3-hydroxy fatty acids (20–90C) called mycolic acids (Sec.
2E.2). Such surface coats are found in relatively few bacterial species, but in plants
waxy surface layers are common, and they contain long-chain (up to 30C) saturated
acids (Sec. 2E.1).

2B   Acyl lipids

Acyl lipids are major constituents of plant and microbial membranes and are often
used as food stores in plants. They comprise a wide variety of structures; most
contain fatty acids as *O*-acyl esters, though sometimes the fatty acids are present as
N-acyl esters. The various types of acyl lipids can be divided into groups according
to the compounds they are combined with. The most common acyl lipids are those

in which fatty acids are esterified to an alcohol, usually glycerol, or less frequently an amino alcohol such as sphingosine. Note, in addition, that some acyl lipids (e.g. plasmalogens) are also ether lipids (see Sec. 2F).

2B.1   Acylglycerols

These contain glycerol and one, two or three esterified fatty acids (Fig. 2.2). Note that not only are positional isomers possible (Fig. 2.2), but the presence of three hydroxyl groups in glycerol produces optical isomerism.

Plant triacylglycerols are of exceptional agricultural and economic importance since they are the 'fats' that everyone is familiar with. In fungi, also, triacylglycerols are often the most abundant type of lipid, sometimes comprising 90% of the total. However, the amount varies considerably with the stage of development and growth conditions. In complete contrast, bacteria do not normally accumulate acylglycerols nor any other 'fats'. The monoacyl and diacyl derivatives of glycerol are important metabolites in all these organisms and are present in small but detectable amounts.

2B.2   Glycerophospholipids

Where the sn-3 position of glycerol contains a phosphate residue, then a glycerophospholipid results. The types of fatty acid in the sn-1 and sn-2 positions often have a characteristic distribution. In bacteria, as in animals, the pattern is usually sn-1-saturated, sn-2-unsaturated. In cyanobacteria acylation is governed by chain length rather than unsaturation, with sn-1-18C, sn-2-16C being the rule. Plants, in contrast, do not show any consistent pattern of acylation, there being differences between species and organelles. One of the hydroxyls of the phosphate can then be further esterified to produce various derivatives as listed in Figure 2.3. These lipids are amphipathic molecules, having a hydrophobic fatty acid domain and a hydrophilic phosphate (+ substituent) domain. The ionisable 'head-group' of the lipid contains a negative charge from the phosphate. However, this may be

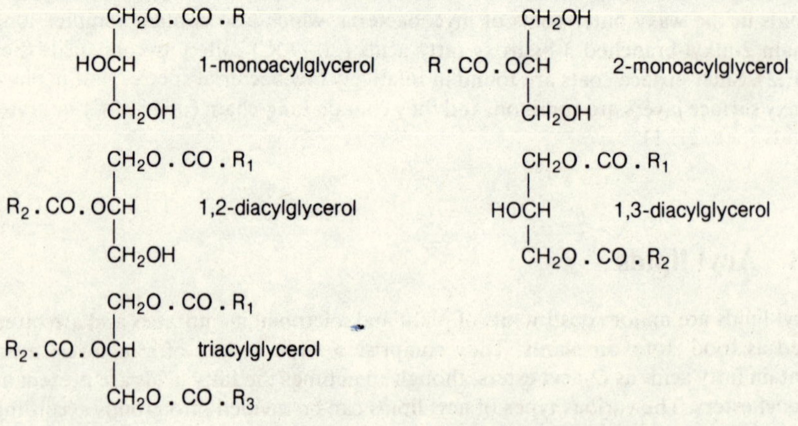

$$CH_2O.CO.R$$
$$|$$
$$HOCH \qquad \text{1-monoacylglycerol}$$
$$|$$
$$CH_2OH$$

$$CH_2OH$$
$$|$$
$$R.CO.OCH \qquad \text{2-monoacylglycerol}$$
$$|$$
$$CH_2OH$$

$$CH_2O.CO.R_1$$
$$|$$
$$R_2.CO.OCH \qquad \text{1,2-diacylglycerol}$$
$$|$$
$$CH_2OH$$

$$CH_2O.CO.R_1$$
$$|$$
$$HOCH \qquad \text{1,3-diacylglycerol}$$
$$|$$
$$CH_2O.CO.R_2$$

$$CH_2O.CO.R_1$$
$$|$$
$$R_2.CO.OCH \qquad \text{triacylglycerol}$$
$$|$$
$$CH_2O.CO.R_3$$

Figure 2.2   Neutral glycerides.

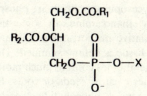

$$CH_2O.CO.R_1$$
$$R_2.CO.OCH \quad O$$
$$CH_2O—P—O—X \qquad \text{general formula}$$
$$O^-$$

substituent	phospholipid	remarks
X = H	phosphatidic acid	negatively charged lipid; important metabolic intermediate, only occurring in trace amounts
X = serine	phosphatidylserine	negatively charged lipid; serine is the L-isomer; widespread but minor lipid
X = ethanolamine	phosphatidylethanolamine	has a net neutral charge at physiological pH; widespread and major lipid
X = choline	phosphatidylcholine	has a net neutral charge; widespread and major lipid in eukaryotes
X = glycerol	phosphatidylglycerol	negatively charged lipid; head group glycerol has sn-1 configuration; widespread and major lipid
X = inositol	phosphatidylinositol	negatively charged lipid; inositol is the myo-isomer; widespread and usually minor lipid
X = phosphatidylglycerol	diphosphatidylglycerol (cardiolipin)	negatively charged lipid; common in bacteria; localised in the inner mitochondrial membrane of eukaryotes

Figure 2.3 The structure and distribution of important phosphoglycerides. In the general formula shown, R_1 and R_2 represent long-chain fatty acids.

balanced by a positive charge on the attached 'X' group. The overall charge on the head-group, together with the amphipathic property, have important implications as regards the ability of glycerophospholipids to form bilayer structures in membranes (see Sec. 6A.1).

An alternative structure to the diacylglycerophospholipids is that found in plasmalogens (see Fig. 2.14). These molecules contain an ether-linked residue at the sn-1 position and an ester at position 2. The ether-linked residue may be saturated, or unsaturated at the carbon atom adjacent to the ether link. Such lipids are common in animals, but have not been reported in plants and apart from some notable exceptions like rumen bacteria are rare in bacteria; they are dealt with in Section 2F.

In plant tissues the most common glycerophospholipids are phosphatidylcholine, phosphatidylglycerol and phosphatidylethanolamine. In some tissues phosphatidylinositol is also important. The relative proportions of these lipids will vary within a species, depending on which tissue is being examined. Thus, leaves with their high content of photosynthetic membranes contain much more phosphatidylglycerol than non-photosynthetic tissues such as seeds or roots. For this reason, also, the green algae (e.g. *Chlorella*) contain phosphatidylglycerol as by far the major glycerophospholipid, whereas brown multicellular algae (e.g. *Fucus*), which have many non-photosynthetic cells, contain equivalent amounts of phosphatidylcholine and phosphatidylethanolamine. Many seeds also contain significant amounts of N-acylphosphatidylethanolamine. In contrast, cyanobacteria (e.g. *Anabaena* spp.) possess a uniquely simple glycerophospholipid composition – only phosphatidylglycerol is present.

Glycerophospholipids are usually the major lipids of bacterial and fungal membranes. In fungi, the major types of glycerophospholipid are the same as for higher plants, although phosphatidylinositol is relatively more important especially in yeasts. Some other structures, such as the phosphorylated derivatives of phosphatidylinositol, have also been reported, but they do not appear to have a widespread occurrence.

In bacteria, phosphatidylglycerol has a wide distribution and is present in all types except the actinomycetes (e.g. mycobacteria), perhaps because of its involvement in cell wall biosynthesis (Sec. 4B.2.2). Sometimes the 3-hydroxyl of the glycerol head-group may be esterified with an amino acid (usually lysine, ornithine or alanine) to form an aminoacylphosphatidylglycerol. These derivatives are more common in Gram-positive species. Less commonly the 2'-hydroxyl of the glycerol head-group may be linked glycosidically to glucosamine or the 3'-hydroxyl group can be acylated.

Diphosphatidylglycerol usually occurs together with phosphatidylglycerol, the relative proportions varying according to growth conditions. Diphosphatidylglycerol is a major lipid in bacteria, in contrast to eukaryotic organisms where it is only present in the inner mitochondrial membrane.

Phosphatidylethanolamine is generally the major glycerophospholipid in Gram-negative bacteria and is a major component of some Gram-positive genera such as *Baccillus*. In a few genera (e.g. some *Clostridium* spp.) the ethanolamine head-group may be methylated to give the mono- and dimethyl derivatives and N-acylphosphatidylethanolamine has been reported in a few bacteria. However, in contrast to higher organisms, the fully methylated derivative, phosphatidylcholine, is rarely a major lipid in bacteria.

Phosphatidylinositol is uncommon in bacteria and is confined to a few Gram-positive genera only. In certain cases mannosides of phosphatidylinositol may also be present (e.g. in actinomycetes). The other glycerophospholipids shown in Figure 2.3 (i.e. phosphatidylserine and phosphatidic acid) are widely distributed in all bacterial, fungal and plant species but are only present in small amounts. Their role as metabolic intermediates is discussed in Section 4B.2.

2B.3 Glycosylglycerides

Some glycerides contain a carbohydrate moiety attached to the sn-3 position of glycerol. Although these glycosylglycerides are present in small quantities in a few animal tissues, and fungal and bacterial species, they are really characteristic of the photosynthetic membranes of higher plants, algae and cyanobacteria. Two galactose-containing lipids are common – diacylgalactosylglycerol and diacyl-digalactosylglycerol (Fig. 2.4). These molecules represent up to 40% of the dry weight of photosynthetic membranes and are consequently the most prevalent membrane lipids in the world.

The two galactose units are linked $\alpha,1 \to 6$ in diacyldigalactosylglycerol and higher homologues with additional galactose units also occur in small amounts in some plants. In certain algae, particularly aquatic species, some of the glycosyl-glycerides contain glucose instead of galactose. In comparison to the glycerophos-pholipids, these galactosylglycerides are less polar because they do not contain any ionisable groups. The properties of these molecules are discussed in Section 6A.1.

The commonest bacterial glycolipids are diacyldiglycosylglycerols but, unlike plants and algae, they do not form such a large proportion of the total lipid. They are found more frequently in Gram-positives but several Gram-negatives, particularly photosynthetic ones, do contain them, usually in smaller amounts than in Gram-positives. The disaccharide moiety is most commonly two glucose, two galactose or two mannose residues linked $\alpha,1 \to 2$ or $\beta,1 \to 6$. Diacyl-digalactosylglycerol (containing the plant-type $\alpha,1 \to 6$ link) is less common; it is important in green photosynthetic bacteria (Sec. 3B.1.4), and in *Treponema* spp. it is the most abundant lipid.

Members of the genus *Butyrovibrio*, which are important anaerobic bacteria in the rumen, contain several unusual galactolipids based upon a diacyl-galactosylglycerol structure. The galactose residue is usually esterified with a short-chain fatty acid (e.g. butyric), and the lipid is in the plasmalogenic form (see Secs 2B.2 & 2F). The diacylgalactosylglycerol may be combined with an (acyl)phosphatidylglycerol (also in plasmalogenic form) through a long-chain fatty acid known as diabolic acid, a 32C dicarboxylic acid with a vicinal dimethyl substitution in the centre of the acyl chain. The resulting molecule (Fig. 2.4) is a phosphoglycolipid (see below), and may span the membrane.

Another important group of bacterial glycolipids are the phosphoglycolipids found in a number of Gram-positive organisms. There are four types: phosphatidylglycolipids confined to N group streptococci; the sn-glycerol-3-phosphoglycolipids of mycoplasmas; the sn-glycerol-1-phosphoglycolipids (related to lipoteichoic acid, Sec. 2E.4); and sn-glycero-1-phosphophosphatidylglycolipids of several Gram-positive genera (Fig. 2.4).

Although not glycosylglycerides, it is appropriate at this point to mention some other bacterial glycolipids. These include acylated sugars such as acylglucose of mycobacterial 'cord factor', the rhamnolipid of *Pseudomonas aeruginosa* and a diacylated glucose attached to D-glyceric acid in the cell envelope of *Nocardia otidis-caviarum* (Fig. 2.4). Further variants are the mycobacterial mycosides (Sec. 2E.2) which are glycosides of *p*-phenol with normal and branched-chain fatty acids esterified to the oligosaccharide. These glycolipids are involved in the pathogenicity of mycobacteria (Sec. 6C.3).

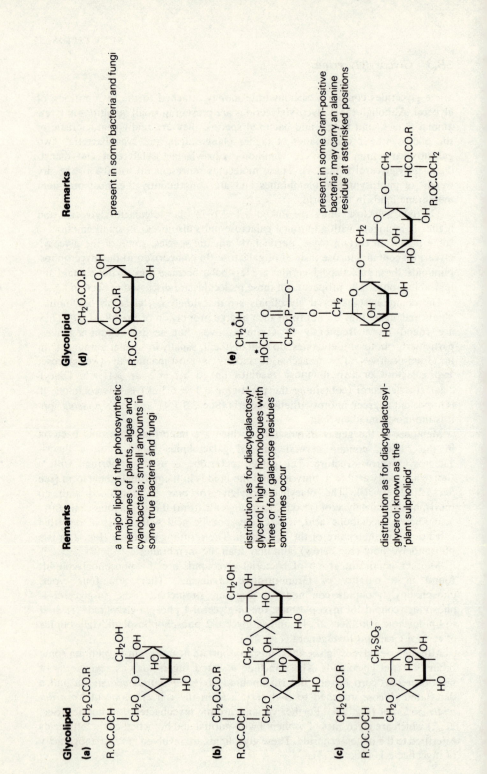

Glycolipid — **Remarks**

(a) a major lipid of the photosynthetic membranes of plants, algae and cyanobacteria; small amounts in some true bacteria and fungi

(b) distribution as for diacylgalactosyl-glycerol; higher homologues with three or four galactose residues sometimes occur

(c) distribution as for diacylgalactosyl-glycerol; known as the 'plant sulpholipid'

(d) present in some bacteria and fungi

(e) present in some Gram-positive bacteria; may carry an alanine residue at asterisked positions

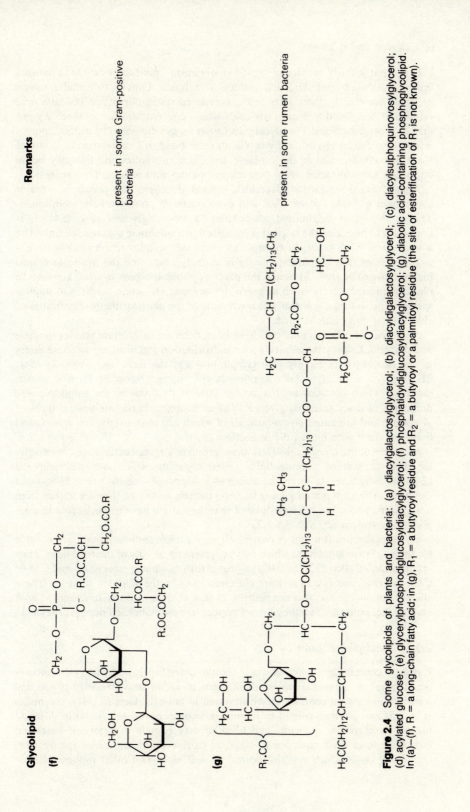

Figure 2.4 Some glycolipids of plants and bacteria: (a) diacylgalactosylglycerol; (b) diacyldigalactosylglycerol; (c) diacylsulphoquinovosylglycerol; (d) acylated glucose; (e) glyceraldiglucosyldiacylglycerol; (f) phosphatidyldiglucosyldiacylglycerol; (g) diabolic acid-containing phosphoglycolipid. In (a)–(f), R = a long-chain fatty acid; in (g), R_1 = a butyroyl residue and R_2 = a butyroyl or a palmitoyl residue (the site of esterification of R_1 is not known).

Glycolipid | Remarks

(f) present in some Gram-positive bacteria

(g) present in some rumen bacteria

Fungi, particularly yeasts and yeast-like organisms, produce a variety of unusual lipids, mostly extracellular, that include glycolipids. Other extracellular fungal products are free hydroxy fatty acids, acetylated sphingolipids, polyol fatty acid esters and acetylated hydroxy fatty acids whose concentrations may reach 2 g per litre of growth medium! The glycolipids (apart from sphingolipids) usually contain a fatty acid linked glycosidically or via an ester bond to a carbohydrate. Some of these glycolipids, such as the ustilagic acids, are antibiotics and this may be an important survival factor when they are competing with bacteria for nutrients.

Apart from the galactosylglycerides, a third glycolipid also occurs as a major component of higher plant, algal and cyanobacterial photosynthetic membranes. This is the plant sulpholipid, diacylsulphoquinovosylglycerol (Fig. 2.4). It is important to note that the sulphur is contained in a sulphonic acid residue linked by a very stable C–S bond, in contrast to other sulpholipids which usually have a sulphate ester. The sulphonate group is strongly acidic and the molecule is thus highly charged *in vivo*. Although the plant sulpholipid occurs in small amounts in photosynthetic bacteria (such as green non-sulphur and some purple non-sulphur species) and in fungi, it is really characteristic of the photosynthetic membranes of chloroplasts and cyanobacteria.

A number of other sulpholipids have been detected in different species of algae and bacteria. Usually these have a sparse distribution and most are sulphate esters of the carbohydrate moiety in a glycolipid – e.g. the trehalose mycolates (Sec. 2E.2) of mycobacteria. An exception is the thermoacidophile *Bacillus acido-caldarius* which contains approximately 10% of its lipids as the sulphonic acid derivative of diacylgalactosylglycerol. The archaebacteria contain several types of glycolipid and sulphated glycolipids, all of which are phytanylglycerol ethers, and these are dealt with more fully in Section 2F.

Members of the *Cytophaga–Flexibacter* group of gliding bacteria are noteworthy for their content of sulphonolipids called capnoids, which are either capnine (2-amino-3-hydroxy-15-methylhexadecane-1-sulphonic acid) or *N*-acylated capnines. Although not *all* gliding bacteria contain capnoids, they are absent from non-gliders, and it has been postulated that they might have a special role in their gliding movement (cf. Sec. 3A.1.2).

Some freshwater (but not seawater) algae, e.g. *Ochromonas danica*, are notable for their chlorosulpholipids, which may represent up to 10% of the total lipid. They are chlorinated alkyl (22C or 24C) sulphates with a sulphate ester at C1 and C14 or C15; the chlorines (up to six) are clustered close to the sulphate residues. These chlorosulpholipids are also noteworthy in that they have polar functions at *both* ends of the molecule, and they do not appear to be metabolised once synthesised.

2B.4 Acylsphingosines

Lipids containing sphingosine (*trans*-D-erythro-1,3-dihydroxy-2-amino-4-octadecene) or a related amino alcohol are of minor importance in plants and micro-organisms, in contrast to the situation in animals. Cerebrosides, ceramides (Fig. 2.5) and phytoglycolipid or related substances have been found in different plants. Cerebrosides in plants and bacteria may contain a variety of bases but only glucose as their component sugar. Phytoglycolipids are minor sphingosine-containing lipids which contain inositol as well as several other monosaccharide

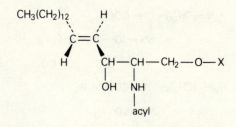

substituent	sphingosine-lipid
X = H	ceramide (acylated sphingosine)
X = sugar	cerebroside
X = phosphocholine	sphingomyelin
X = phosphoinositol	ceramide phosphoinositol (present in yeast)

Figure 2.5 Structures of some sphingosine-based acyl lipids.

residues. Several variants of the basic structure have been identified from different plant sources.

The diatom *Nitzschia alba* contains a sulphonolipid (1-deoxyceramide-1-sulphonic acid) that is similar to the capnoids of gliding bacteria except that it is based on sphingosine rather than capnine (see above). Interestingly, this diatom also moves by gliding.

Most fungi seem to contain small amounts of the usual sphingolipids. The most thorough studies have been with various filamentous fungi such as the infamous *Amanita muscaria*, in which ceramides and cerebrosides represent about 1% of the mycelial dry weight. A number of unusual sphingolipids have been detected in different species, including several inositol-containing phosphorylceramides in baker's yeast where they make up 39% of the inositol-containing lipids.

In bacteria, sphingolipids are seldom found. They have been most often reported in anaerobic bacteria such as *Bacteroides* spp., with ceramides predominating. Several aminolipids containing 15–18C alkylamines (i.e. similar to sphingosine but lacking hydroxyl groups) are major lipids in *Deinococcus* spp., and may be characteristic of this newly designated group of radiation-resistant bacteria.

2B.5 Acylornithines

Representatives of a wide range of bacterial types, including photosynthetics, pathogens and sulphur-oxidisers, contain amino lipids that are based not on the amino alcohol sphingosine but on the amino acid ornithine. The structure of these so-called 'ornithine lipids' varies from one organism to another (see Fig. 2.6 for some representative structures). The ornithine α-amino group is amide-linked to a fatty acid, which can be a 2- or 3-hydroxy fatty acid that is further esterified via its hydroxyl group to a second fatty acid (Fig. 2.6a). The α-carboxyl group may be esterified with a fatty alcohol (Fig. 2.6b), the hydroxyl group of a 3-hydroxy fatty acid, or a short diol that is further esterified to a fatty acid (Fig 2.6c).

The ornithine lipids do not contain phosphorus, and it is not known what their function is. They are amphipathic (like phospholipids) but are positively charged

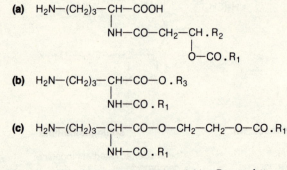

Figure 2.6 Some examples of bacterial ornithine lipids. R_1 = a fatty acid of variable structure, depending on the species of bacterium; R_2 = a 3-hydroxy fatty acid; R_3 = a fatty alcohol.

(unlike phospholipids) and can form membranes. An ornithine lipid (Fig. 2.6a) is the major lipid in *Desulphovibrio gigas,* and in *Pseudomonas rubescens* this ornithine lipid replaces phospholipid as the main lipid in phosphate-limited cultures.

2C Terpenoids

Plants produce an amazing variety of chemicals based on a branched 5C building block (isoprene unit). Some of these are primary metabolites such as sterols and carotenoids. Others may contribute to important structures as in the side-chains of chlorophyll or in enzyme prosthetic groups. However, the majority of the terpenoids synthesised by plants are secondary metabolites and are uniquely plant products.

Terpenoids are traditionally divided into classes based on the number of carbon atoms they contain (or, in some cases, were derived from). Their nomenclature is shown in Table 2.2, together with important examples of each class, and some structures are given in Figure 2.7. Since not all members of a given class necessarily contain the same number of carbon atoms, we will need to use the generic class names (hemiterpenes, etc.).

The simplest terpenoids, the hemiterpenes, occur in very small amounts in most plants and algae. Isoprene occurs abundantly as a volatile emission from leaves, while certain primative plants, such as bracken fern, may produce especially large amounts. Traditionally, the monoterpenoids (10C) have been thought of as components of essential oils, being accumulated by only a few oil-bearing plants (e.g. *Citrus* spp., mint, sage and pine) with specialised secretory structures.

The sesquiterpenes (15C) are the largest single class of terpenoid and their structural diversity is remarkable. Like the monoterpenes, the accumulation of large quantities of sesquiterpenes in plants is generally associated with specialised oil glands. However, several members of this class play well defined and essential roles in both higher and lower plants. For example, the plant hormone abscisic acid,

Table 2.2 Terpenoid classes.

Class	Usual carbon no.	Remarks
hemiterpenes	5C	small amounts only in plants; comparatively rare in algae
monoterpenes	10C	widely distributed in plants (e.g. important in Pinaceae); some cyclohexanoid glycoside derivatives; uncommon in algae and fungi
sesquiterpenes	15C	abscisic acid and furanoid phytoalexins are plant examples; fungal antibiotics
diterpenes	20C	plant examples include gibberellins and gymnosperm resin acids; widespread occurrence in fungi, e.g. gibberellins
sesterterpenes	25C	rare in plants and fungi; plant pathogenic fungi are main source
triterpenes	30C	plant examples include saponins, sterols, squalene and components of resins, latex and cuticle; sterols are common in fungi; small amounts of squalene, but not sterols, are present in bacteria; hopanoids are found in some bacteria and plants; tetrahymenol in *Tetrahymena*
polyterpenes	$(5C)_n$	plant examples include rubber, gutta and chicle (the traditional chewing-gum base); carotenoids are widely distributed in plants, algae, fungi and bacteria; plasto-quinones and ubiquinones are common; dolichols are found in plants and some fungi; undecaprenol is present in bacteria

the furanoid phytoalexins, numerous fungal antibiotics, and distasteful compounds such as lactone anti-feedants are all sesquiterpenes.

Gibberellins are examples of diterpenes (20C). They were originally discovered as phytotoxins produced by a plant pathogenic fungus. Such fungi have developed the ability to manipulate the physiology of their host plant by synthesising compounds which then act as plant hormones. To date over 50 gibberellins are known, although it is not clear how many are physiologically important.

Open-chain terpenoid alcohols and their derivatives are referred to as prenols. Several electron-transport quinones (plastoquinones, menaquinones and ubiquinones) contain prenyl side-chains. These molecules are further named from the number of isoprene (5C) units in their side-chain – e.g. ubiquinone$_{10}$ (UQ–10) contains ten such units. The most common bacterial quinones are MQ–8 and UQ–8 and UQ–9, which are often partially hydrogenated in the isoprenoid side-chain. Plastoquinone is found in plants and eukaryotic algae, but not in photosynthetic bacteria. The various quinones have different redox potentials but serve essentially similar functions as mobile electron carriers in membranes.

A family of *cis*-polyprenols, called dolichols, of varying chain length (80–100C) is found in plants and animals. These lipophilic molecules are important as carriers of sugar residues in cell wall synthesis and protein glycosylation, facilitating the passage of a hydrophilic molecule across the hydrophobic membrane barrier. In bacteria, the corresponding component is a 55C isoprenoid alcohol, called undecaprenol or bactoprenol. Its pyrophosphate derivative serves a similar function as a lipophilic sugar carrier in the biosynthesis of peptidoglycan,

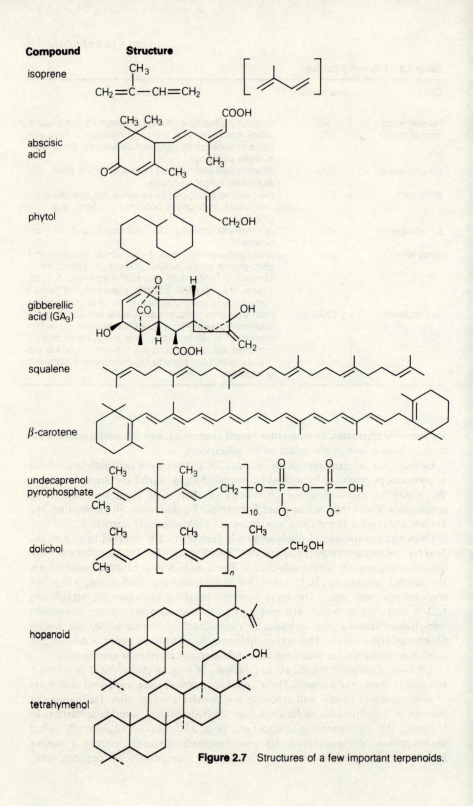

Figure 2.7 Structures of a few important terpenoids.

lipopolysaccharide O-antigens, teichoic acid, teichuronic acid and several other extracellular polysaccharides.

An important role for terpenoids in all photosynthetic organisms is as a constituent of the prenyl side-chain of chlorophylls. The several chlorophylls are magnesium porphyrins which have slightly different ring structures and side-chains (Fig. 2.8). Bacteriochlorophylls are distinguished by the presence of

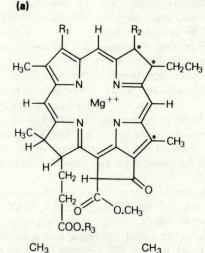

(a)

$R_3 = HOCH_2-CH=\overset{\overset{\displaystyle CH_3}{|}}{C}-CH_2(-CH_2-CH_2-\overset{\overset{\displaystyle CH_3}{|}}{CH}-CH_2-)_3H,$ phytol

$R_3 = HO(-CH_2-CH=\overset{\overset{\displaystyle CH_3}{|}}{C}-CH_2-)_4H,$ geranylgeraniol

$R_3 = HO(-CH_2-CH=\overset{\overset{\displaystyle CH_3}{|}}{C}-CH_2-)_3H,$ farnesol

(b)

chlorophyll type	R_1	R_2	R_3	absorption max. (nm)
chlorophyll a	$-CH=CH_2$	$-CH_3$	phytol	663
chlorophyll b	$-CH=CH_2$	$-CHO$	phytol	645
chlorophyll c	$-CH=CH_2$	$-CH_3$	H	631
chlorophyll d	$-CHO$	$-CH_3$	phytol	688
bacteriochlorophyll a	$-C(CH_3)O$	$-CH_3$	geranylgeraniol or phytol	770
bacteriochlorophyll b	$-C(CH_3)O$	$-CH_3$	geranylgeraniol or phytol	795

Figure 2.8 (a) The basic chlorophyll structure. R_3 may be one of the three prenyl side-chains shown. Substituents at asterisked positions may also vary in different bacterio-chlorophylls and chlorophylls. (b) Tabulation of the possible R_1 and R_2 side-groups shown in the general structure. Variation of R_1, R_2 and R_3 leads to the different chlorophylls and bacteriochlorophylls, with their characteristic absorption maxima. The fact that chlorophyll c does not contain a prenyl side-chain, but has instead a hydrogen atom at that position makes this chlorophyll water-soluble rather than lipid-soluble. Bacteriochlorophylls c, d and e contain farnesol, and also differ in their substituents at the three asterisked positions.

geranylgeraniol or farnesol instead of phytol (Fig. 2.8). These variations give the molecules slightly different absorption maxima, especially in the red region of the spectrum. With the exception of chlorophyll *c* all of the chlorophyll pigments contain a prenyl side-chain and this allows that part of the molecule to be anchored in the bilayer of the photosynthetic membrane where it appears to be tightly associated with particular proteins. The binding in these pigment–protein complexes is very strong, and allows chlorophylls to be separated on sodium dodecyl sulphate–polyacrylamide gels in association with their protein components! In addition, chlorophylls *in vivo* show multiple absorption bands which are believed to be caused by small differences in their environment. These differences are very important in ensuring that the absorption of light energy is so efficient (approximately 98%) in photosynthesis.

The distinction between triterpenes and steroids (Sec. 2D) is not clear-cut. Traditionally, 30C isoprenoid compounds are considered to be triterpenes whereas structures based on the tetracylic cyclopentanoperhydrophenanthrene ring structure are sterols. Many of the free triterpenes which are components of resins, latex or cuticle are powerful surfactants – e.g. they can cause lysis of blood cells. Some of them have been used as fish poisons by primitive tribes. Triterpenes also include gossypol from the cotton plant and linonin, an extremely bitter compound found in lemons.

Carotenoids are pigmented terpenes which usually contain eight isoprene units (40C), comprised of two 20C halves joined 'head to head'; they have more than nine conjugated double bonds, which makes them brightly coloured. The hydrocarbon carotenoids are called carotenes, and those containing oxygen are called xanthophylls. There are over 300 naturally occurring carotenoids and these are widely distributed in higher plants, bacteria, algae, animals and fungi. Those of higher plants, algae and fungi are compared in Table 2.3.

The carotenoid composition of photosynthetic bacteria has been used as a basis for classifying them, as they contain some characteristic compounds. For example, methoxylated carotenoids are found in most photosynthetic bacteria, while the green and purple sulphur and non-sulphur genera can be distinguished by the presence of acyclic carotenoids (e.g. spirilloxanthin) in purple bacteria and aromatic carotenoids in green bacteria.

By contrast the occurrence of carotenoids in non-photosynthetic bacteria is random and, as in plants, closely related species may have quite different colours. Hence, they are not generally useful for taxonomic purposes, although there are a number of 30C, 40C and 50C carotenoids that are unique to non-photosynthetic bacteria. For example, a 30C carotenoid isolated from *Staphylococcus aureus* is termed 'bacterial phytoene'. The 50C terminally hydroxlated carotenoid, bacteriorubin, found in extremely halophilic bacteria, is responsible for the red colour of some salt lakes and contaminated salted fish or hides. Carotenoid glycosides, together with the acyclic 40C carotenoids having terminal oxygen functions and the hydroxylated or glycosylated 50C carotenoids, are commonly found in bacteria (and cyanobacteria), but are rare or absent from eukaryotes. β-Carotene and other carotenes are less widely distributed in non-photosynthetic bacteria than in fungi, and the xanthophylls are generally different from those occurring in higher plants.

Many fungi contain carotenoids but their distribution is random; levels may vary

Table 2.3 Major pigments in higher plants, algae and fungi.

	α-Carotene	β-Carotene	Lutein	Zeaxanthin	Violoxanthin	Neoxanthin	Fucoxanthin	Others
higher plants (photosynthetic tissues)	±	+	+	±	+	+	−	−
green algae (Chlorophyta)	+	+	+	+	+	+	−	−
red algae (Rhodophyta)	+	+	+	+	−	−	−	−
yellow-green algae	−	+	−	−	−	−	+	−
brown algae (Phaeophyta)	−	+	−	−	+	−	+	−
golden-brown algae	+	+	−	−	−	−	+	−
Cryptophyta	−	+	−	+	−	+	−	+
Euglenophyta	±	+	−	+	−	+	−	+
fungi	−	+	−	±	−	−	−	+[a]

[a] γ-Carotene is particularly common.

considerably from species to species, and are also dependent on growth conditions. γ-Carotene is particularly common, while α-carotene has not been detected (Table 2.3). Fungal xanthophylls are often carboxylic acids such as torularhodin, and higher plant or algal xanthophylls are not found in fungi.

Algae all contain β-carotene, but other major carotenoids vary from genus to genus, giving them their distinctive pigmentation (Table 2.3). Lichens (which consist of a symbiotic association of an alga with a fungus) usually contain a mixture of carotenoids synthesised by both organisms although, in some cases, the fungus may make no contribution.

As mentioned above, the photosynthetic tissues of higher plants accumulate a characteristic carotenoid mixture (Table 2.3). In addition, roots, seeds, fruits and flower petals may accumulate large amounts of carotenoids to give the familiar delightfully bright colours. Examples of each would be carrots, maize, tomatoes and many flowers respectively.

The polyterpenes are high molecular weight compounds. These hydrocarbons, of basic formula $(C_5H_8)_n$, include rubber (of molecular weight $\sim 10^6$) and chicle, the traditional chewing gum base. Polyterpenes are produced by a wide variety of species but only a few are commercially important. Indeed, the advent of synthetics has largely led to the eclipse of gutta as an important natural product.

2D   Sterols

Sterols represent one group of triterpenoids. They are derived from squalene, a 30C compound, and contain a fused four-ring system (Fig. 2.9). The major sterols found in plants and algae, viz. sitosterol ($\sim 70\%$), stigmasterol ($\sim 20\%$), campesterol ($\sim 5\%$) and cholesterol ($\sim 5\%$) (Fig. 2.9), differ mainly in the nature of the side-chain – which has different degrees of substitution and unsaturation. In addition to free sterols, plants also contain sterol esters (where a fatty acid is esterified to the 3-hydroxyl), sterol glycosides (where the 3-hydroxyl of the sterol forms a glycosidic linkage with the 1-position of a hexose, usually glucose) and acylated sterol glycosides (where the 6-position of the hexose is esterified with a fatty acid).

Depending on the plant tissue, the relative abundance of sterols, sterol glycosides, sterol esters and acylated sterol glycosides can vary. Any of these classes can predominate. Furthermore, the nature of the fatty acids in sterol esters can also vary with any one of the commonly occurring plant fatty acids being present. However, in acylated sterol glycosides, palmitate and linoleate predominate.

Sterols are important in many fungi and some species have been used extensively as experimental organisms for investigations of sterol biosynthesis. Yeasts accumulate the greatest amounts of sterols, which may represent up to 10% of the cellular dry weight! Phycomycetes frequently contain cholesterol, but ergosterol is usually the major sterol, representing over 90% of the sterol in some fungi such as *Mucor* spp. Again, in various yeasts and mushrooms ergosterol is the most abundant sterol. The sterols of rust fungi differ considerably from other fungi in that there is an absence of ergosterol and a predominence of 29C sterols. The Pythiaceae are unusual in that they cannot make sterols and can grow in their

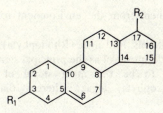

basic structure

Sterol (R₁ = OH)

cholesterol

campesterol

sitosterol

stigmasterol

R₂ substituent

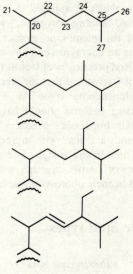

Sterol derivative

sterol ester
(e.g. palmitoyl ester)

sterol glucoside

acylated sterol glucoside
(e.g. palmitoyl ester)

R₁ substituent

Figure 2.9 Commonly occurring plant and algal sterols and their derivatives.

absence, but they must acquire them from the environment for reproduction (asexual or sexual).

Algae produce a wide variety of sterols. Red algae (Rhodophyta) may synthesise cholesterol, 24-ethylcholesterol or desmosterol as the principal sterol according to the species. Fucosterol appears to be the main sterol of golden-brown (Chrysophyta) and brown (Phaeophyta) algae, whereas diatoms produce brassicasterol in large amounts.

Bacteria characteristically lack sterols in their membranes, the notable exceptions being some methylotrophs, in which squalene and a 4-methyl sterol are present. Sterol synthesis is blocked at the latter compound which does not accumulate in plants or animals. A few bacteria, e.g. *Pseudomonas testosteroni*, contain steroid-binding proteins, which concentrate extracellular steroids specifically at the membrane surface for transport into the cell and ultimate degradation as an energy/carbon source in the cytoplasm.

The uncyclised precursors of sterols (e.g. squalene) are more frequently found in bacteria and it is assumed that the sterol ring-forming cyclisation reactions, which are oxygen-dependent, arose at a later stage in evolution. Why some methane-utilising bacteria should have developed this ability is curious. Some bacteria contain hopanoids, products of the cyclisation of unoxidised squalene; these compounds are dealt with in Section 4C.5.

For a long time it was thought that sterols were absent from cyanobacteria, but small amounts of sterols, together with the terpenoids phytol and squalene, are usually found in their photosynthetic lamellae.

2E Other lipid types

2E.1 Cutin, suberin and waxes

Plants are unique in that the structural component of their outer surface (cuticle) is made up of a hydroxy fatty acid polymer – cutin. In contrast, other organisms utilise polymers of amino acids or carbohydrates. Underground parts and the surfaces of wounds in plants are covered by another type of lipid-derived polymeric material – suberin. These polymers are associated with or embedded in a complex mixture of less polar lipids called waxes. Except in certain unusual species (e.g. the desert plant, jojoba) waxes are not found in the internal organs of plants but are located at the surface. However, in certain algae (the marine zooplankton are a notable example) waxes are used extensively as an energy store.

The major components of plant waxes are shown in Table 2.4. It should be emphasised that the term 'wax' is not a specific one but derives from the similarity of properties of plant waxes to those of beeswax. In general, most of the major wax components are non-polar molecules with long hydrocarbon chains. Whether the accumulated molecules are of odd-chain or even-chain length depends on their pathway of synthesis (Sec. 4D.1.1).

In both cutin and suberin 16C and 18C components are common. But, whereas cutin is characterised by dihydroxy fatty acids, suberin contains ω-hydroxy acids and dicarboxylic fatty acids as major components. Both complexes contain phenolic constituents and these are especially important in suberin. The main compositional

Table 2.4 The major components of plant waxes.

Compound	Structure	Occurrence
n-alkanes	$CH_3(CH_2)_nCH_3$	most plants; major components usually 29C or 31C
iso-alkanes	$CH_3CH(CH_3)R$	not as widespread as n-alkanes; usually 27C, 29C, 31C and 33C
ketones	$R_1 . CO . R_2$	not as common as alkanes; usually 29C and 31C
secondary alcohols	$R_1 . CH(OH)R_2$	as common as ketones; usually 29C and 31C
	$CH_3CH(OH)R$	uncommon in cuticular waxes; more common in suberin waxes; odd-chain length, 9–15C
β-diketones	$R_1 . COCH_2CO . R_2$	usually minor, but in some species (e.g. barley) may be major components; mainly 29C, 31C and 33C
primary alcohols	$R . CH_2OH$	most plants; even-chains predominate, usually 26C and 28C
acids	$R . COOH$	very common; even-chains predominate, usually 24–28C

Abbreviations: R, R_1 and R_2 are alkyl chains.

differences between cutin and suberin are listed in Table 2.5. In any individual plant, the composition of the surface cutin varies somewhat in different parts of the plant. In general, the more slowly growing tissues such as fruits or stems are covered in longer chain length components than the rapidly growing parts, e.g. leaves. Although, as mentioned above, cutin and suberin are essentially external coverings, they may be found surrounding specialised structures such as the seed embryo, or lining organs such as nectaries or the juice sacs of citrus fruits.

The cell walls of fungi contain appreciable quantities of aliphatic hydrocarbons. They are similar in chain length to those of higher plants and algae, and are assumed to play a similar functional role.

In comparison with plants, bacteria do not contain large amounts of waxy materials in their surface layers. The mycobacteria are an important exception and they are dealt with in Section 2E.2. A few genera, notably *Acinetobacter* spp. and the related *Micrococcus cryophilus* (a Gram-negative coccus, despite its name), contain wax esters (up to 13% of the total lipid), simple esters of a fatty acid and a fatty alcohol, which have similar compositions to the fatty acids of other acyl lipids in the same organisms. These wax esters are not part of an external protective coating, but serve as membrane components or energy stores.

Hydrocarbons have been found in a wide range of bacteria, but often in trace

Table 2.5 The main compositional differences between cutin and suberin.

Monomer	Cutin	Suberin
dicarboxylic acids	minor	major
in-chain-substituted acids	major	minor (sometimes substantial)
phenolics	low	high
very-long-chain (20–26C) acids	rare and minor	common and substantial
very-long-chain alcohols	rare and minor	common and substantial

Reproduced with permission from Kolattukudy, P. E. 1980. In *Biochemistry of Plants*, Vol. 4, P. K. Stumpf & E. E. Conn (eds), 591. New York: Academic Press.

amounts only. Micrococci may contain large amounts (up to 20% of the total lipid). The hydrocarbon composition usually reflects that of the other acyl lipids, so that micrococci have branched hydrocarbons. Their function is unknown.

2E.2 Mycobacterial cell wall lipids

Mycobacteria, as well as the related nocardiae and corynebacteria, contain large amounts of lipid compared with most other bacteria. This lipid is largely in the cell wall and, in addition to the more common bacterial lipids, is typified by a bewildering array of high molecular weight and complex waxy molecules. The latter contain sugars and peptides as well as long-chain highly substituted fatty acids, many of which are unique to this group of bacteria. The lipids are important in several aspects of the pathogenicity of these organisms (Sec. 6C.3). It is for this reason that they are also important historically and in fact were studied in detail before most 'conventional' membrane lipids of other bacteria.

The mycobacterial cell wall contains three main components: a skeleton composed of arabinogalactan mycolate, covalently linked through a phosphodiester bond to peptidoglycan; free lipids that can be extracted with solvents; and peptides

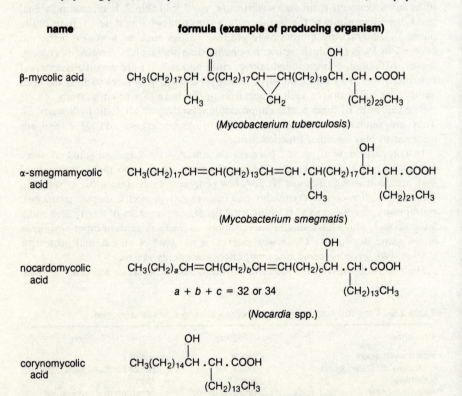

Figure 2.10 Some major mycolic acids found in mycobacteria and related actinomycetes.

which are removed only by proteolysis. Some strains also contain a glucan component.

The arabinogalactan mycolate is a highly branched polymer of D-arabinose and D-galactose (in a 5 : 2 ratio) in which about every tenth arabinose contains a mycolic acid esterified to the 5′-hydroxyl. Mycolic acids are 2-branched, 3-hydroxy long-chain (22–90C) fatty acids, which may be (di)unsaturated or contain methyl branches, methoxy groups and cyclopropane rings (Fig. 2.10). The 2-branch is usually a normal 22–24C acyl chain. Similar arabinogalactan mycolates are present in cell walls of *Nocardia* and *Corynebacterium* species. Generally, the mycolic acids of mycobacteria are of larger molecular weight than those of corynebacteria and nocardiae (but see Sec. 3A.1.1).

(a)

(b)

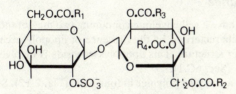

(c)

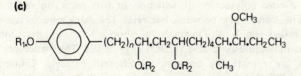

(d)

R.O — Phe — *allo*Thr — Ala — alaninol — sugar
                              |
                         O — sugar

Figure 2.11 Structures of mycobacterial lipids: (a) trehalose mycolate, where R_1 and R_2 are mycolic acids (see Fig. 2.10); (b) sulpholipid, where R_1–R_4 are palmitic acid or very-long-chain branched fatty acids; (c) mycosides A and B, where R_1 is a mono- or a trisaccharide and R_2 is a 12–18C saturated fatty acid or a mycocerosic acid (e.g. $CH_3(CH_2)_{21}(CH(CH_3)$-$CH_2)_2CH(CH_3)COOH$); (d) mycosides C, where R is a long-chain (28–34C) methoxy fatty acid (e.g. $CH_3(CH_2)_{22}CH=CHCH(OCH_3)CH_2COOH$), Phe = phenylalanine, *allo*Thr = *allo*-threonine and Ala = alanine.

The extractable lipids account for 25–30% of the weight of the cell wall and consist of wax D, cord factors, mycosides and sulpholipids. The wax D is an autolysis product of the skeleton where the peptidoglycan has been partially cleaved. The cord factors are so named because they were extracted from pathogenic mycobacteria (e.g. tubercle bacilli) that grow in the form of 'cords' on the surface of solid media. In spite of the name, these factors are not responsible for this distinctive mode of growth. Cord factor is a mixture of trehalose 6,6'-mycolates (Fig. 2.11). Trehalose is a disaccharide of glucose joined 'head to head' (C1→1'). The constituent mycolic acids are essentially the same as those found in wax D, and vary according to the producing organism.

The trehalose mycolate may be sulphated at the 2-position and further acyl groups added at several positions on both sugar rings of the trehalose to produce sulpholipids (Fig. 2.11). Instead of mycolic acids (as in cord factor), the sulpholipids have a mixture of palmitic acid and very-long-chain (31–46C) fatty acids, with up to ten methyl branches, known as phthioceranic acids.

The outermost layer of the mycobacterial cell wall contains mycosides, which are either phenolic glycolipids (mycosides A and B) or glycolipid peptide (mycoside C) containing unusual methylated sugars such as methylated fucose and rhamnose (Fig. 2.11). The mycosides C have a 3-oxygenated (hydroxy or methoxy) fatty acid (approximately 28C) amide-linked to phenylalanine which is the N-terminal amino acid of a variety of short oligopeptides to which the sugars are attached.

2E.3 Lipopolysaccharide

Gram-negative bacteria have a cell envelope containing two membranes (Fig. 1.1); the outer membrane is characterised by the presence of lipopolysaccharide in the outer leaflet of its bilayer structure. This lipopolysaccharide is involved in several aspects of pathogenicity (Sec. 6C.1).

Lipopolysaccharide is a complex polymer in four parts (Fig. 2.12). Starting from the outside of the cell, there is a polysaccharide of very variable structure which carries several antigenic determinants and is often called the O-antigen. This is attached to a core polysaccharide which is in two parts, an outer core and a backbone. The cores vary between bacteria. The backbone is connected to a glycolipid, called lipid A, through a short 'link' usually composed of 3-deoxy-D-mannooctulosonic acid (KDO) (Fig. 2.12). The structures of the link and lipid A are remarkably constant in different bacteria. Consequently the presence of KDO is often used as a marker for lipopolysaccharide (or outer membrane), although KDO is not present in all bacterial lipopolysaccharides.

Lipid A consists of a disaccharide of glucosamines that are highly substituted with phosphate, fatty acids and KDO. The amino groups are substituted exclusively with 3-hydroxymyristate while the remaining hydroxyl groups are acylated with a mixture of 12C, 14C and 16C saturated fatty acids and 3-myristoxymyristate. There is some microheterogeneity between bacteria with regard to the nature of the fatty acids and also the amount of phosphate or type of sugar. The hydroxy fatty acids most commonly have the 3-D configuration.

The fatty acids are sometimes referred to as 'bound fatty acids' and lipopolysaccharide as 'bound lipid' since they are not released by extraction procedures used for phospholipids, etc. They are only released after saponification

(a)

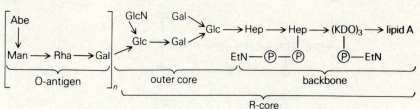

(b)

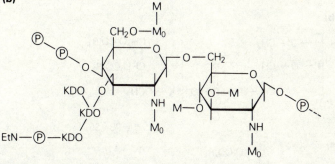

Figure 2.12 Generalised structures of (a) lipopolysaccharide and (b) lipid A, showing the KDO link region. Abe = abequose, Man = mannose, Rha = rhamnose, Gal = galactose, Glc = glucose, Hep = heptose, KDO = deoxy-D-mannooctulosonic acid, $\textcircled{P}$ = phosphate, EtN = ethanolamine, M = myristrate and M_0 = β-hydroxymyristrate.

of whole cells or purified lipopolysaccharide (obtained by hot phenol–water extraction of cells). The fatty acid composition of lipopolysaccharide is characterised not only by the presence of its hydroxy fatty acids but also by the absence of unsaturated or cyclopropane fatty acids so typical of the extractable acyl lipids of Gram-negatives.

2E.4   Lipoteichoic acids and other anionic polymers

The cell walls and membranes of most Gram-positive bacteria contain a series of highly anionic polymers. Quantitatively one of the most important is teichoic acid, which is a polymer of glycerol 1-phosphate or ribitol phosphate. Some of the *sn*-2 hydroxyl groups are either glycosylated or esterified with D-alanine (Fig. 2.13). The membrane teichoic acids are based on glycerol 1-phosphate and a proportion are covalently linked to a glycolipid, such as diacyldiglycosylglycerol, to give a lipoteichoic acid (Fig. 2.13). The amount of glycosylation and esterification, and the types of sugars or glycolipid vary amongst bacteria. In some (e.g. streptococci and lactobacilli) lipoteichoic acids are group-specific antigens, and variations in glycosylation form the basis of group typing.

In streptococci, the lipoteichoic acid may be acylated on the glucose attached to the diacylglycerol. These bacteria also contain (acylated) diacyldiglucosylglycerols, which are probably the biosynthetic precursors of membrane lipoteichoic acid (Sec. 4D.4).

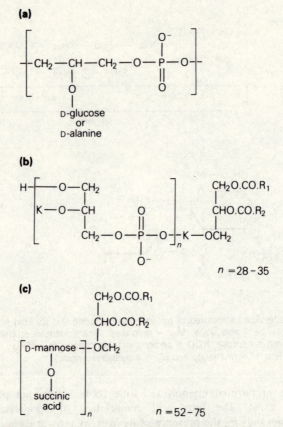

(a)

(b)

$n = 28-35$

(c)

$n = 52-75$

Figure 2.13 Structures of some anionic polymers in bacteria: (a) teichoic acid (general formula); (b) lipoteichoic acid (from *Streptococcus lactis*); (c) lipomannan (from *Micrococcus lysodeikticus*). K = kojibiose (6-*O*-β-D-glucosyl-D-glucose); R_1 and R_2 = fatty acids

Another type of anionic polymer found in Gram-positive bacteria such as *Micrococcus lysodeikticus* is succinylated lipomannan (Fig. 2.13). The mannose residues are joined in a mixture of 1→2, 1→3 and 1→6 linkages, and up to 25% are substituted with succinic acid, which gives the polymer its negative charge. Like teichoic acid, the succinylated lipomannan is anchored in the membrane by its diacyglycerol moiety.

Actinomycetes, which also lack lipoteichoic acid, contain instead a complex heteropolysaccharide, composed mainly of mannose, glucose and galactose substituted with fatty acids, glycerophosphate, lysine and alanine. Its detailed structure and relationship to the cell membrane are not known.

2F Ether lipids

There are three main types of ether lipid in bacteria. Anaerobic bacteria, including those of the rumen, contain relatively large amounts of plasmalogens (Fig. 2.14)

(a)

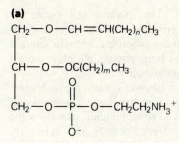

$$CH_2-O-CH=CH(CH_2)_nCH_3$$
$$CH-O-OC(CH_2)_mCH_3$$
$$CH_2-O-\overset{\displaystyle O}{\underset{\displaystyle O^-}{P}}-O-CH_2CH_2NH_3^+$$

(b)

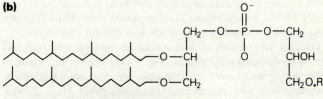

$R = HPO_3^-$ = phosphatidylglycerophosphate
$R = H$ = phosphatidylglycerol
$R = SO_3^-$ = phosphatidylglycerosulphate

(c)

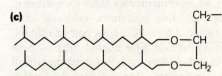

$$CH_2-O-glucose-mannose-galactose-O.R$$

$R = SO_3^-$ = glycolipid sulphate
$R = H$ = triglycosyl ether

(d)

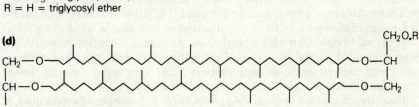

$R = CH_2OH.CHOH.CH_2O.PO_3^-$ = phosphoglycolipid
$R = H$ = diglycosyldibiphytanyldiglycerol tetraether

(e)

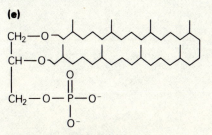

macrocyclic glycerol ether

Figure 2.14 Structures of bacterial ether lipids: (a) anaerobic bacterial plasmalogens; (b)–(e) archaebacterial phytanyl ethers. In (d), the sugars are usually glucose and galactose.

which have both ester and ether linkages. The plasmalogen forms of phosphatidylethanolamine and phosphatidylglycerol and its derivatives are the most important. Very few aerobic or facultatively anaerobic bacteria contain plasmalogens. A few species of higher plants and fungi contain the plasmalogen forms of phosphatidylethanolamine and phosphatidylcholine.

Some species of gliding bacteria, e.g. *Myxococcus xanthus*, contain plasmalogens. Others, in particular *Stigmatella aurantiaca*, contain large amounts of non-plasmalogenic dialkyl and monoalkyl-monoacyl phospholipids, as well as the more usual ester-linked diacyl phospholipids. Very recently the extremely thermophilic, anaerobic, sulphate-reducing bacterium *Thermodesulfotobacterium commune* has also been shown to contain largely diether-linked phospholipids and glycolipids with mainly *br*-17 : 0 alkyl chains. This organism is found in the same extreme environment as some archaebacteria, which also contain ether lipids.

In contrast, the glycerol ether lipids of archaebacteria contain only phytanyl (saturated 20C isoprenoid-derived) alkyl chains (Fig. 2.14). The various kinds of ether include diphytanylglycerol diether analogues of phosphatidylglycerol, phosphatidylglycerolphosphate and phosphatidylglycerolsulphate (Fig. 2.14b), which are the major lipids in extreme halophiles, as well as a number of glycosylated diphytanylglycerol diethers (Fig. 2.14c) and dibiphytanyldiglycerol tetraethers (Fig. 2.14d). The tetraethers are the predominant lipids in thermoacidophiles and thermoalkaliphiles, whereas methanogens have a mixture of di- and tetraethers. The thermoacidophiles also contain a calditolglycerol tetraether lipid in which one glycerol is replaced by a 9C polyol named calditol. Some of the phytanyl chains also contain up to four 5-membered rings on each chain. Most recently, in studies of a new extremely thermophilic (optimum growth temperature, 85 °C!) methanogenic archaebacterium, *Methanococcus jannaschii*, which was isolated from 'black smoker' sediment at the East Pacific Rise hydrothermal vents, a novel macrocyclic ether (Fig. 2.14e) has been found in the polar membrane lipids. Interestingly, this kind of structure, in which the glycerol oxygens are bridged by the biphytanyl chain, was one of those first proposed, and later discounted, for the tetraethers. The significance of these and other variations for membrane structure in archaebacteria, and adaptation to their hostile environments, are discussed in Sections 6A.1 and 6A.2.

Recently, an unusual lipid (diacylglycerol-*O*-(*N*,*N*,*N*-trimethyl)homoserine), in which the homoserine residue is ether-linked to glycerol, has been found in several algae, including species of the halotolerant genus *Dunaliella*.

3

Distribution of lipids

3A Lipid distributions in different organisms and their use in taxonomy

3A.1 *Lipid composition in bacterial taxonomy*

3A.1.1 Introduction

Bacteria are classified into genera and species primarily on the basis of their colonial appearance, cellular morphology and growth characteristics (Sec. 1B). This information is usually supplemented with other biochemical features, including lipid composition. We shall show that not all aspects of bacterial lipids are generally useful in this respect, although there are many examples of the specific taxonomic use of lipid composition (and metabolism). It must be remembered, particularly when considering quantitative differences, that the cultural conditions may influence markedly the relative amounts of some lipids (Secs 3E & 6A.2). In addition fatty acids may be taken up from the growth medium and incorporated into acyl lipids with or without further modification.

A broad distinction can be made between the lipid composition of Gram-positive and Gram-negative genera. The major Gram-positive lipids are phosphatidylglycerol and phosphatidylethanolamine, in which the acyl chains are predominantly odd-numbered branched-chain fatty acids; Gram-positives may also contain lipoteichoic acid in their membranes. In contrast, Gram-negatives have phosphatidylethanolamine with even-numbered straight-chain and cyclopropane acyl chains; the cell envelope also contains lipopolysaccharide.

Apart from broad generalisations such as these, specific lipid compositions are used to aid in more detailed classification – e.g. to assign particular species to a genus. Quite often anomalies are discovered in the lipid composition of a species, whose presence in a genus has been questioned already on other grounds. In this way lipid composition complements other taxonomic determinants. For example, a case has been put forward for restricting the genus *Corynebacterium* to human and animal pathogens (and some related saprophytic strains) based on a variety of morphological and biochemical criteria. This proposal is supported by lipid analyses of true corynebacteria, which have low molecular weight mycolic acids (22–36C), straight-chain and monounsaturated fatty acids, and dihydrogenated menaquinones with eight and/or nine isoprene units. Lipid analysis was used to show that *Corynebacterium paurometabolum* is not a typical coryneform: it has a fatty acid and phospholipid composition similar to mycobacteria and nocardiae, but it differs from them in having unsaturated menaquinones and long (68–76C), highly unsaturated (two to six double bonds) mycolic acids. Analysis of mycolic acids has been particularly useful in the taxonomy of mycobacteria and related organisms; for example, the presence of long-chain mycolic acids in *Mycobacterium leprae* helped to define this important pathogen.

Perhaps the best example of the taxonomic use of lipid composition is in the definition of the new ur-kingdom, the archaebacteria (Sec. 1B). Despite the fact that the archaebacteria consist of groups with several distinct phenotypes – viz. methanogens, extreme halophiles, and extreme thermoalkaliphiles and thermoacidophiles – they all contain phytanyl ether-linked lipids (Sec. 2F).

3A.1.2   Fatty acid composition

Bacterial fatty acid composition correlates well with the Gram stain, apart from a few clear exceptions; these include the Gram-positive lactobacilli and some clostridia which have a 'Gram-negative-like' fatty acid composition, and some Gram-negative *Bacteroides* and *Fusibacterium* spp., which contain branched-chain fatty acids. Fatty acid composition can also be used to make finer taxonomic distinctions. Even for a large and diverse family such as the Enterobacteriaceae a good separation of genera and species can be achieved by fatty acid compositional analysis, especially when this is performed using modern computer techniques for detecting similar patterns and groups. In the genus *Bacillus*, two groups can be distinguished: the '*B. subtilis* group' containing predominantly *anteiso*-branched 15 : 0 and the '*B. cereus* group' containing *iso*-branched 15 : 0. Occasionally a specific fatty acid may characterise a genus – e.g. 13C fatty acids in *Cellulomonas* spp. Gliding bacteria all have a very unusual fatty acid composition: *Chloroflexus* and *Hepetosiphon* spp. have polyunsaturated fatty acids and *Myxococcus*, *Cytophaga* and *Flexibacter* spp. contain hydroxy branched fatty acids (in their phospholipids as well as lipopolysaccharide). In each case their unusual fatty acids may be an adaptation to provide flexibility for movement.

3A.1.3   Neutral lipid composition

The hydrocarbon composition of a species is usually simpler than the fatty acid composition, but there is little consistent taxonomic pattern. The main taxonomic use of hydrocarbons is in helping to differentiate micrococci (in which hydrocarbons may comprise up to 20% of the total lipid) from staphylococci (which lack hydrocarbons). The two species *Sarcina lutea* and *Sarcina flava* were found to contain the same hydrocarbons, and have been reclassified on the basis of this and other criteria as the single species *Micrococcus luteus*.

Wax esters are seldom found in bacteria, but their presence in *Acinetobacter* spp. may be taxonomically significant.

In contrast, all bacteria contain menaquinones and/or ubiquinones, and the length of the isoprenoid side-chain and its state of reduction have been used to distinguish species (see Sec. 3A.1.1).

The limited usefulness of carotenoids in bacterial classification is discussed in Section 2C.

3A.1.4   Phospholipid composition

Little use can be made of phospholipid composition in bacterial taxonomy because the major phospholipids (phosphatidylglycerol and phosphatidylethanolamine) are

Table 3.1 A comparison of the lipid compositions and metabolisms of some spirochaete genera.

Lipid characteristic	Spirochaeta	Spirochaete genus Treponema	Leptospira
diacylgalactosylglycerol	+++	+++	−
phosphatidylcholine	−	+++	−
phosphatidylethanolamine	−	++	+++
β-oxidation	−	−	+++
fatty acid synthesis	+++[a]	±[b]	+[c]
fatty acid desaturation	−	−	+++

Abbreviations: +++, present in all or most species; ++, present in some species; +, present in few species; −, present in no species.
[a]Synthesise saturated and unsaturated fatty acids by the anaerobic pathway.
[b]Generally, free-living species synthesise fatty acids (by the anaerobic pathway), whereas parasitic species cannot and depend on their host for a supply of fatty acids.
[c]Saturated fatty acid synthesis only.
Modified with the permission of the American Society of Microbiology from Livermore, B. P. and R. C. Johnson 1974. *J. Bacteriol.* **120**, 1269–73.

common to the majority of species. Even the distribution of phosphatidylcholine, which is relatively rare in bacteria, does not follow any clear taxonomic pattern (but see Table 3.1). The same is true of phosphatidylinositol, although phosphatidylinositol mannosides are characteristically the major phospholipids of actinomycetes.

A recent example of the use of phospholipid composition in taxonomy concerns the group of radiation-resistant micrococci. On the basis of features such as their cell wall structure, 16s RNA sequence and the presence of 16 : 1 rather than 15 : 0 *br* as their major fatty acid, they have been assigned to a new genus, *Deinococcus*. Interestingly, it is now becoming apparent that deinococci have an extremely unusual phospholipid composition, lacking *all* of the common phospholipids found in most bacteria! Instead, most of their major acyl lipids are (phospho)glycolipids containing alkylamines as unique constituents (see Sec. 2B.4).

3A.1.5 Glycolipid composition
The commonest glycolipids in bacteria, particularly Gram-positives, are glycosylglycerides and despite the variety of sugar residues some genera have characteristic combinations. Thus staphylococci have diacyldiglucosylglycerol but streptococci contain diacylglucosyglycerol as well as diacyldiglucosylglycerol.

Gram-negatives either lack glycolipids or contain only small amounts. Instead, they possess lipopolysaccharide, which has a similar structure in all Gram-negatives and consequently is of little taxonomic value. Small variations in the O-antigen polysaccharide (Sec. 2E.3) form the basis of phage-typing of, for example, *Salmonella* strains.

3A.1.6 Lipid metabolism
Little taxonomic use is made of the metabolism of lipids, simply because this aspect of lipid biochemistry has been studied in too few organisms. Most is known about fatty acid metabolism and it is possible to divide bacteria into three broad groups on

this basis. The first (e.g. some bacilli) contains branched fatty acids and has no unsaturated fatty acids and thus no oxygen requirement. The second (e.g. *Escherichia* and *Clostridium* spp.) contains straight-chain fatty acids and uses the anaerobic pathway of unsaturated fatty acid biosynthesis (see Sec. 4A.2) and so has no oxygen requirement. The third (e.g. *Pseudomonas* spp.) also has straight-chain fatty acids, but uses aerobic desaturation to synthesise unsaturated fatty acids and so requires oxygen.

The ability to oxidise fatty acids or even to use them as sole carbon and energy source is invoked as a taxonomic criterion – e.g. for *Acinetobacter* and *Leptospira* spp. Several aspects of fatty acid metabolism, as well as their phospholipid and glycolipid composition, help to distinguish various spirochaete genera (Table 3.1). *Leptospira* spp. are the only spirochaetes that are aerobes; although few species synthesise fatty acids, most can desaturate exogenous saturated fatty acids. All other spirochaetes are obligate or facultative anaerobes and either synthesise both saturated and unsaturated fatty acids by the anaerobic pathway or none at all (Table 3.1).

3A.2   *Distribution of lipids in photosynthetic organisms*

3A.2.1   *Introduction*
Four major groups of photosynthetic organisms can be distinguished, namely higher plants, eukaryotic algae, cyanobacteria (prokaryotic algae) and photosynthetic bacteria. The latter are divided into purple (sulphur and non-sulphur) and green photosynthetic bacteria. Plants, eukaryotic algae and cyanobacteria carry out oxygenic photosynthesis, whereas bacteria carry out anoxygenic photosynthesis. The photosynthetic membranes of oxygen-evolving organisms also contain different lipids from those of the photosynthetic bacteria (Table 3.2). In particular the former do not contain much phospholipid but, instead, are rich in glycosylglycerides (Sec. 3B); the small amount of phospholipid is mainly phosphatidylglycerol. In contrast, the lipid of the photosynthetic membranes of bacteria is mainly phospholipid and, perhaps significantly, phosphatidylglycerol is a major component.

Table 3.2   Lipid composition of photosynthetic membranes.

	PC	PG	PE	DPG	OL	MGDG	DGDG	SQDG
higher plants	trace	+	–	–	–	+	+	+
algae	–	+	–	–	–	+	+	+
cyanobacteria	–	+	–	–	–	+	+	+
green bacteria	–	+	–	+	–	+	–	+
purple sulphur bacteria	–	+	+	+	–	+	+	–
purple non-sulphur bacteria	+	+	+	+	+	–	–	+

The table is a simplification and not *all* members of a group may contain all of the lipids listed.
Abbreviations: PC, phosphatidylcholine; PG, phosphatidylglycerol; PE, phosphatidylethanolamine; DPG, diphosphatidylglycerol (cardiolipin); OL, ornithine-containing lipids; MGDG, diacylgalactosylglycerol; DGDG, diacyldigalactosylglycerol; SQDG, diacylsulphoquinovosylglycerol, the 'plant sulpholipid'.

3A.2.2   Classification of plants

Over a quarter of a million flowering plants have been classified as well as over 100 000 lower plants including 10 000 algae. The classification of such a range is, obviously, extremely difficult and has been based mainly on morphological considerations. However, now that accumulated metabolites can be identified conveniently, plants can be screened rapidly and analytical data can contribute to the solution of taxonomic problems. Sometimes chemical and morphological characteristics have evolved together. For example, plants which are classified morphologically as Flacourtiaceae accumulate unusual cyclopentenoid fatty acids in their seeds. On the other hand, plants from completely different families may accumulate the same unusual compounds. For instance, this happens in seeds which contain conjugated ethylenic fatty acids. Moreover, some plant lipids are of such wide occurrence that their taxonomic value is negligible and the external environment can influence composition to quite a high degree. The taxonomic significance of fatty acid composition, in particular, must be treated with reserve.

In higher plants, the fatty acid, acyl lipid and pigment composition of whole leaves is remarkably similar and offers no taxonomic possibilities. However, the cuticular waxes show considerable variation with similar compositions in related species. This has classification potential which has not, so far, been exploited. In contrast, the diverse fatty acid contents of seed oils have been utilised to classify plant families.

Taxonomy based on seed oil composition depends primarily on the principal component fatty acids. Although this does not always lead to clearly defined groups nor to unique grouping for a particular botanical family, most families can be classified in this way (Table 3.3).

The system suffers from two major problems. Firstly, it is dependent on accurate analysis to determine the relative proportions of different acids, many of which

Table 3.3 Examples of the use of seed oil composition for classifying plants.

Stored fatty acids	Fatty acid example	Plant family example
(i) common fatty acids[a] only		
linolenate-rich		Linaceae
myristate/laurate-rich		Palmae
(ii) common acids[a] + characteristic minor acids[b]		
saturated acids	$20:0, 22:0$	Leguminosae
unsaturated acids	$\omega 6\text{-}18:3$	Boraginaceae
(iii) common acids[a] + characteristic unusual acids		
non-conjugated ethylenic acids	$18:1(11c)$	Asclepidaceae
conjugated ethylenic acids	$12t$-family	Bignoniaceae
acetylenic acids	$18:1(6a)$	Simaroubaceae
substituted acids	epoxy acids	Valerianaceae
branched chain acids	cyclopentenoid acids	Flacourtiaceae

[a]Common acids: $12:0, 14:0, 16:0, 18:0, 18:1(9c), 18:2(9c,12c), 18:3(9c,12c,15c)$.
[b]Minor acids: $6:0, 8:0, 10:0, 17:0, 20:0, 22:0, 24:0, 16:1(9c), 22:1(13c), 16:3(7c,10c,13c), 18:3(6c,9c,12c), 18:4(6c,9c,12c,15c), 20:4(5c,8c,11c,14c), 22:6(4c,7c,10c,13c,16c,19c)$.

occur as major components in most seeds. These proportions are to some extent also influenced by growth conditions. Secondly, some species that are closely related yield seed oils which are very different and vice versa. This may be because seed oil composition is not particularly important to the plant, so long as the lipid can be used efficiently during germination. Thus, there are no strong factors which ensure that seed compositions are maintained during evolution, and as new strains of plants are developed exceptions to the classification become more frequent.

3A.2.3 Classification of algae

Algal lipids contain ordinary fatty acids typical of higher plants but, in addition, are often a good source of 'minor' polyunsaturated fatty acids. Classification of algae on this basis suffers from some of the same problems as those for seeds. That is,

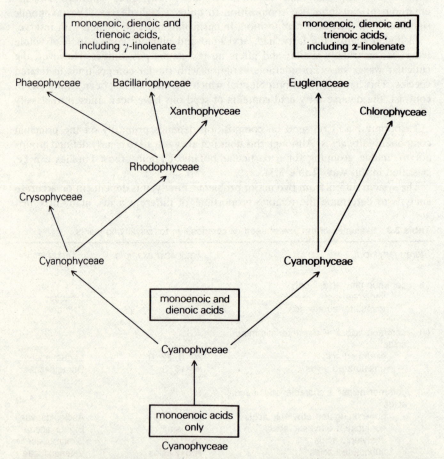

Figure 3.1 Classification of algae based on their fatty acid composition and metabolism. The relationships of the different groups of algae are shown, with their characteristic fatty acids given in boxes. The arrows do *not* imply phylogenetic relationships. [Redrawn with permission from Hitchcock, C., and B. W. Nichols 1971. *Plant lipid biochemistry*, p. 32. London: Academic Press. © Academic Press Inc. (London) Ltd.]

different algae contain the same compounds and reliance must be placed on quantitative concentrations of the same acids which may vary considerably according to cultural conditions.

Some generalisations, however, are apparent. Green freshwater algae grown photoautotrophically contain a fatty acid composition similar to green leaves, except that the proportion of α-linolenate is somewhat lower. Heterotrophic growth leads to a large reduction in polyunsaturated fatty acids. Marine algae (including green Chlorophyceae, red Rhodophyceae, brown Phaeophyceae and diatoms or Bacillariophyceae) synthesise considerable quantities of 20C and 22C polyunsaturated acids in contrast to freshwater algae. Cyanobacteria contain little or no polyunsaturated fatty acid and no $trans$-Δ3-hexadecenoic acid (Sec. 3B).

Another criterion which can be used in classification is the ability of an alga to synthesise α- or γ-linolenic acid. The former acid is typical of higher plants and green algae, whereas γ-linolenic acid is associated with animals, ciliates and amoebae. Some algae accumulate linolenate by both pathways, their relative activity being affected by growth conditions. These differences in fatty acid synthesis (and hence occurrence) can be used for classification purposes (Fig. 3.1).

Although acyl lipid constituents can help in the classification of algae, the pigment composition has been the most useful aid to taxonomy. Thus the carotenoid content has led to the recognition of the green, brown, red and other algal groups. These broad groups of algae can now be further subdivided by reference to the detailed description of pigment composition (see Table 2.3), although again, just as with divisions based on fatty acid composition, the relative proportions of carotenoids can be affected by growth conditions.

3B Subcellular distribution of lipids

3B.1 Lipid composition of membranes

3B.1.1 Introduction
Lipids are not evenly distributed throughout cells. This is particularly true of eukaryotic cells with their diversity of intracellular membranes compared with prokaryotic cells. Indeed, so specific are some localisations that the presence (or absence) of certain lipids may be used in much the same way as enzyme markers to assess the purity of subcellular functions. The major lipid components of almost all plant and microbial membranes are glycerolipids (some exceptions will be mentioned later). A summary of the major features of higher plant, algal, yeast and bacterial membrane lipids is given in Table 3.4.

3B.1.2 Cytoplasmic (plasma) membranes
Approximately half of the dry weight of plant plasma membranes is lipid. Major components include phospholipids (up to 65%), glycolipids (up to 20%), sterols (up to 5%) and neutral lipids, including hydrocarbons, diacylglycerols and pigments. The main phospholipids are phosphatidylcholine, phosphatidylethanolamine and phosphatidylinositol. Sterols, such as stigmasterol and sitosterol (which together represent about 80% of the total sterol of most plants) are not usually quantitatively as important as cholesterol is in animal plasma membranes. Both the

Table 3.4 The major lipids of plant and microbial membranes.

Organism and membrane	Major lipids	Major fatty acids
(a) plants		
chloroplast lamellae	MGDG > DGDG > SQDG (70%) PG (10%) chl a > chl b	$18:3 \gg 16:0$
extrachloroplastic membranes	PC > PE > PI = PG (90%)	$16:0 = 18:2 > 18:3 > 18:1$
(b) bacteria		
cytoplasmic membranes		
Gram-positive	PG > DPG = PE > aminoacyl-PG glycolipids	$15:0br > 17:0br$
Gram-negative	PE > PG > DPG	$16:0 > 16:1 = 18:1$
	lipopolysaccharide[a]	$OH-14:0$
photosynthetic membranes		
purple bacteria	PE > PG > DPG = PC bchl a > bchl b	$18:1 > 16:1 = 16:0$
green bacteria	PG, DPG MGDG, glycolipid-II, SQDG bchl c > bchl d	$18:1 > 16:0 = 14:0$
(c) cyanobacteria:		
total membranes	MGDG > DGDG > SQDG (80%) PG only (10%) chl a only	$16:0 = 18:3 > 16:1 = 18:2$
(d) yeast		
total membranes	PC = PI = PE > DPG = PS (80%)	$16:0 = 18:1 > 16:1 = 18:2$
(e) brown algae (seaweeds)		
total membranes	MGDG = DGDG = SQDG (80%) PE = PG = PC > DPG = PI (15%) chl a > chl c	$16:0 = 18:1 = 18:2 = 18:3 = 18:4 = 20:4 = 20:5$

Abbreviations: MGDG, diacylgalactosylglycerol; DGDG, diacyldigalactosylglycerol; SQDG, diacylsulpho-quinovosylglycerol; PG, phosphatidylglycerol; chl a, chlorophyll a; chl b, chlorophyll b; PC, phosphatidyl-choline; PE, phosphatidylethanolamine; PI, phosphatidylinositol; DPG, diphosphatidylglycerol (cardiolipin); bchl, bacteriochlorophyll; PS, phosphatidylserine.
[a] Present in outer membrane only.

amounts and relative distributions of free sterols, sterol esters and (acylated) sterol glycosides vary markedly from one plant to another and between tissues.

As in plants, phosphatidylcholine has now been noted as the major lipid in algal plasma membranes. Together with phosphatidylethanolamine and phosphatidylinositol, phosphatidylcholine is usually found in yeast plasma membrane preparations, which often contain a high proportion (>50%) of neutral lipid (mainly triacylglycerols). However, it is not clear whether triacylglycerol is a true membrane component or, perhaps more likely, is present in vesicles that co-purify with the membranes.

The major lipids of bacterial cytoplasmic membranes are phospholipids, mainly phosphatidylethanolamine and phosphatidylglycerol. In addition to phospholipids, glycolipids are present, particularly in Gram-positives, while Gram-negatives contain lipopolysaccharides (Sec. 2E.3) in their outer membrane (see below). Neutral lipids generally form a small proportion of the lipid in bacterial cytoplasmic membranes, particularly those of Gram-negatives. Hydrocarbons, free fatty acids and diacylglycerol (but not triacylglycerol) have been reported in lipid analyses of some Gram-positives, and a few species are notable in containing large amounts of neutral lipid – e.g. 50% of the total lipid of *Sarcina lutea* is hydrocarbon and acylglycerols. However, in none of these cases is it clear whether neutral lipids are *bona fide* components of any cell membrane.

As already noted, steroids do not usually occur in true bacteria, the exception being some methylotrophs which contain intracytoplasmic membranes (Sec. 3B.1.5), although there is no evidence for a preferential location of the sterols between these membranes and the cytoplasmic membrane. The mycoplasmas, a group of wall-less parasitic prokaryotes, require sterols for growth; these they obtain from the host and incorporate into their plasma membrane.

The so-called 'cell envelope' of Gram-negative bacteria consists of an inner (cytoplasmic or plasma) membrane and an outer membrane; Gram-positives only have a cytoplasmic membrane (see Fig. 1.1). The outer membrane contains a greater proportion (approximately 60% in *Escherichia coli*) of the total cell lipid, because it has a higher lipid : protein ratio than the inner membrane, which contains more enzymes and proteins such as the cytochromes. The inner and outer membranes usually contain the same types of phospholipids, but there is a larger percentage of phosphatidylethanolamine in the outer membrane (Table 3.5).

3B.1.3 Mitochondria

Mitochondria have been prepared from plant tissues by several techniques and differences in these methods may have influenced the somewhat diverse compositions that have been reported (for example, the presence of endogenous phospholipase D may generate large quanitites of phosphatidic acid). In general,

Table 3.5 A comparison of the lipid compositions of cytoplasmic and outer membranes of some Gram-negative bacteria.

| | Phospholipid composition (%, w/w) | | |
	PE	PG	DPG
Escherichia coli			
cytoplasmic membrane	72	24	4
outer membrane	88	10	2
Salmonella typhimurium			
cytoplasmic membrane	60	33	7
outer membrane	81	17	2
Proteus mirabilis			
cytoplasmic membrane	60	10	30
outer membrane	78	4	17

For abbreviations see Table 3.4.

Table 3.6 Lipid compositions of individual plant membranes.

Plant	Membrane	Percentage of total acyl lipids								
		Phospholipids						Glycolipids		
		PC	PE	DPG	PI	PS	PG	MGDG	DGDG	SQDG
potato	plasma membrane	37	41	2	18	—	—	◄————— 2 —————►		
tuber	mitochondrial outer	54	30	—	11	—	5	◄————— trace —————►		
	mitochondrial inner	41	37	15	5	—	3	◄————— trace —————►		
	microsomal	44	32	1	15	2	2	◄————— 6 —————►		
spinach	chloroplast envelope	27	—	—	1	—	8	22	32	5
	chloroplast thylakoid	3	—	—	1	—	9	51	26	7
wheat	chloroplast thylakoid	1	—	—	—	—	9	47	36	7

Plasma membranes and the microsomal fraction also contain significant quantities of sterols; thylakoid membranes contain chlorophyll and carotenoid pigments; chloroplast envelopes contain carotenoid pigments.

The potato membrane fractions were prepared from a variety low in endogenous acyl hydrolase activity. For abbreviations see Table 3.4.

high proportions (up to 98% of the total lipid) of phospholipids are found, with phosphatidylcholine and phosphatidylethanolamine as major components. Phosphatidylylinositol is concentrated in the outer mitochondrial membrane, whereas diphosphatidylglycerol is found exclusively in the inner mitochondrial membrane, of which it is a major component (Table 3.6). In this context it is interesting to note that diphosphatidylglycerol is commonly found in bacterial plasma membranes and it has been suggested that animal mitochondria arose by the engulfment of a primitive prokaryote ('endosymbiont hypothesis'). In other respects there is usually little resemblance between the lipid composition of the inner mitochondrial membrane and bacterial plasma membranes. However, the plasma membrane of *Paracoccus denitrificans* bears some similarity to the inner mitochondrial membrane in that it contains up to 3% phosphatidylcholine (depending on the growth conditions). Because of its 'mitochondria-like' cytochromes, this organism has been proposed as an archetype of that which may have formed the original endosymbiont.

The mitochondrial membranes from certain yeasts, such as *Saccharomyces cerevisiae*, have been isolated and characterised and found to contain predominantly phospholipids. Neutral lipids make up about 15% and sterols about 5% of the total lipids. Yeasts are useful to the biochemist in that they can be used to study the development of mitochondria. The promitochondria, however, contain the same phospholipids as mitochondria (though with decreased amounts of diphosphatidylglycerol and phosphatidylethanolamine) but lack sterol.

3B.1.4 Photosynthetic membranes

Chloroplast membranes are unique in nature in that the major lipid components of their lamellae are glycosylglycerides (Table 3.6) and not phospholipids. The three glycosylglycerides common to these photosynthetic membranes are diacyl-galactosylglycerol, diacyldigalactosylglycerol and diacylsulphoquinovosylglycerol, in order of their abundance. The major phospholipid (about 10% of the total lipid)

Table 3.7 Pigment distribution in photosynthetic membranes.

Organism	Major pigments
higher plants	chlorophyll *a*, chlorophyll *b*, α-carotene, β-carotene, luteol
green algae	chlorophyll *a*, chlorophyll *b*, α-carotene, β-carotene, luteol
brown algae	chlorophyll *a*, chlorophyll *c*, β-carotene, fucoxanthol, luteol
cyanobacteria	chlorophyll *a*, β-carotene, phycocyanins, phycoerythrins, luteol, myxoxanthol
purple bacteria	
non-sulphur	bacteriochlorophyll *a*, bacteriochlorophyll *b*, spirilloxanthol, spheroidene
sulphur	bacteriochlorophyll *a*, spirilloxanthol, okenone
green bacteria	bacteriochlorophyll *c*, bacteriochlorophyll *d*, bacteriochlorophyll *e*,[a] chlorobactene

[a] Only in 'brown-coloured' *Chlorobium* spp.

is phosphatidylglycerol. The surrounding membranes of the chloroplast, the double envelope, have recently been isolated and analysed. The inner envelope membrane contains a similar acyl lipid composition to the lamellae whereas that of the outer envelope membrane resembles endoplasmic reticulum. A striking and experimentally very useful property of the chloroplast membranes is their significant content of different pigment molecules. The occurrence and proportion of these chromophores differs with species but whereas carotenoids are present in both the envelope and the internal lamellae, chlorophylls are found only in the latter. Typical examples of pigment distributions are shown in Table 3.7, which shows that while chlorophyll *a* and chlorophyll *b* are the main chorophylls in higher plants, other derivatives may be the major green pigments in different organisms. The carotenoids are particularly diverse and this may be seen most obviously in some algae or bacteria where the colour results from the carotenoid content (e.g. brown algae with fucoxanthol or *Rhodospirillum rubrum* with spirilloxanthol). In addition, carotenoids, including those of flower petals, fruits such as tomato (containing leucopene), citrus fruits and peppers.

Algae, including the blue-green algae (more properly called cyanobacteria), synthesise and accumulate the major varieties of acyl lipid found in the photosynthetic membranes of higher plants. Thus, the three glycosylglycerides characteristic of chloroplast lamellae are present in major proportions in the various classes of algae. *Euglena gracilis* has frequently been used for studies of lipids because it can be grown heterotrophically in the dark (when it obtains carbon and energy by oxidising organic compounds dissolved in the growth medium) or photoautotrophically in the light (when it obtains energy from light and carbon by 'fixing' atmospheric carbon dioxide). In the dark it makes large amounts of wax esters and phospholipids (phosphatidylcholine and phosphatidylethanolamine). Upon illumination, it forms chloroplasts and, hence, accumulates large amounts of the three glycosylglycerides and phosphatidylglycerol.

Table 3.8 A comparison of major acyl lipids and their fatty acid compositions in chromatophores and non-photosynthetic membranes of *Rhodopseudomonas sphaeroides*.

Membrane type	Acyl lipids (percentage of total)		Major fatty acids (percentage of total)		
			16:0	18:0	18:1
chromatophores	PE	22	6	14	80
	PG	54	6	11	82
	DPG	10	7	14	78
	PC	10	5	12	82
	SQDG	4	22	15	54
non-photosynthetic membranes	PE	28	6	14	78
	PG	48	7	15	77
	DPG	10	10	15	73
	PC	11	8	15	75
	SQDG	2	20	15	61

For abbreviations see Table 3.4.

When purple bacteria are grown under appropriate light and nutritional conditions internal photosynthetic membranes are formed that are functionally equivalent to the thylakoid membranes of higher plant chloroplasts. After breaking the bacteria, these photosynthetic membranes can be isolated as 'chromatophores'. The major lipids of these chromatophores are phospholipids, while the typical plant chloroplastic lipids, glycosylglycerides, are absent. The phospholipid compositions of chromatophores and cytoplasmic membranes are similar, except that chromatophores are relatively enriched in phosphatidylglycerol (Table 3.8). But unlike in plant chloroplast lamellae, chromatophore phosphatidylglycerol is not the sole phospholipid and does not contain *trans*-Δ3-hexadecenoic acid (see Sec. 3B.2.2). Small amounts of the typical plant sulpholipid, diacylsulphoquinovosylglycerol, are present in both chromatophore and cytoplasmic membranes. Chromatophores are rich in bacteriochlorophyll *a* and, occasionally, bacteriochlorophyll *b* (Sec. 2C).

The organisation of the photosynthetic membranes in green sulphur bacteria is somewhat different to that in purple sulphur bacteria (Fig. 3.2 & cf. Fig. 3.3). In contrast to chromatophores, the major lipid in the 'chlorosomes' (or 'chlorobium vesicles') of green sulphur bacteria such as *Chlorobium thiosulphatophilum* is diacylgalactosylglycerol; phospholipids are absent from these vesicles. The chlorosomes have a very high bacteriochlorophyll : protein ratio and, unlike chromatophores of purple bacteria, they may not house the 'reaction centres' which trap light energy during photosynthesis; these are probably in the cytoplasmic membrane. Compared with the chlorosomes the cytoplasmic membrane has a distinctive lipid composition in that it lacks diacylgalactosylglycerol and instead contains so-called glycolipid II (a galactose, rhamnose-diacylglycerol probably) together with phospholipids.

3B.1.5 Microsomes

When eukaryotic cells are fractionated the microsomal fraction contains a mixture of small membrane vesicles. Often it is assumed that 'microsomal' is synonymous

(a)

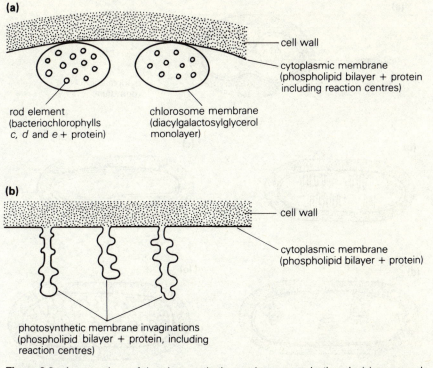

cell wall

cytoplasmic membrane
(phospholipid bilayer + protein
including reaction centres)

rod element
(bacteriochlorophylls
c, d and *e* + protein)

chlorosome membrane
(diacylgalactosylglycerol
monolayer)

(b)

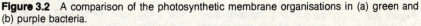

cell wall

cytoplasmic membrane
(phospholipid bilayer + protein)

photosynthetic membrane invaginations
(phospholipid bilayer + protein, including
reaction centres)

Figure 3.2 A comparison of the photosynthetic membrane organisations in (a) green and (b) purple bacteria.

with 'endoplasmic reticulum' – but this is rarely the case! Microsomal fractions from higher plant cells will almost invariably contain plastid envelope, plasma membrane and fragments of other organelles in addition to smooth and rough endoplasmic reticulum. The average lipid composition of plant 'microsomes' tends to resemble that of plasma membranes, except that glycolipids are more prevalent (Table 3.6) and sterols somewhat less so. Of the other higher plant organelles, the membranes of nuclei, glyoxysomes and peroxisomes are rich in the phospholipids, phosphatidylcholine and phosphatidylethanolamine.

Bacteria do not contain internal membranes that are equivalent to those of the endoplasmic reticulum and, for example, carry ribosomes. Indeed, it is often stated that one of the characteristics of prokaryotic organisms is that they do not contain an extensive system of intracellular membrane or membranous organelles. However, this is a misleading statement and most bacteria, particularly Gram-positive species, contain a limited amount of intracellular membrane organised in structures called **mesosomes** (Fig. 3.3). These are often located close to the site of cross wall/membrane formation during cell division and consist of an invagination of the plasma membrane containing a folded chain of vesicles, lamellae or tubes of membrane. No unique function for these organelles has been discovered, despite several claims to the contrary, and indeed their existence has been questioned on the grounds that they are artefacts of the electron microscopic

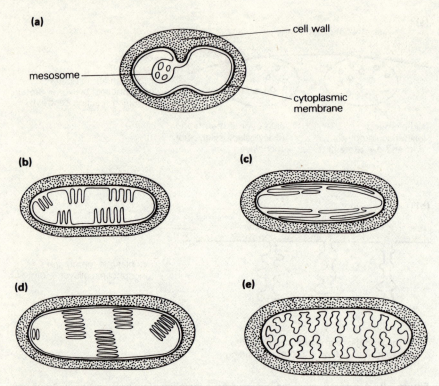

cell wall

mesosome

cytoplasmic membrane

(a)

(b)   (c)

(d)   (e)

Figure 3.3 A comparison of a mesosome with other intracellular membranes of bacteria: (a) a typical mesosome; (b) tubular invaginations; (c) peripheral lamellae; (d) lamellar stacks; (e) vesicular invaginations.

preparation techniques. In view of their location in the cell, it has been suggested that they could be preferential sites of membrane biogenesis. However, in the few investigations which have been reported neither their lipid composition nor rate of lipid biosynthesis are significantly different from that of the plasma membrane. It has been reported that the mesosomes of *Staphylococcus aureus* contain virtually all the membrane lipoteichoic acid (Sec. 2E.4) of the cell, while those of *Micrococcus lysodeikticus* are enriched in succinylated lipomannan (Sec. 2E.4).

Apart from mesosomes, which are widespread amongst bacterial species, and the photosynthetic organelles discussed above, various specialised groups of bacteria are noted for the presence of internal membranes that are usually related to specific functions such as ATP production from H_2 or CH_4 oxidation. Amongst the species that have extensive intracellular membranes are the denitrifying bacteria, methylotrophs and photosynthetic bacteria. These groups possess a wide variety of membrane systems, including lamellae and vesicles which in some cases almost fill the cell (Fig. 3.3). For example, when the bacterium *Azotobacter vinelandii* is grown on methane as the sole carbon source the synthesis of a vast internal membrane system is induced in which resides the nitrogen-fixing enzymes. These membranes are enriched in phosphatidylethanolamine and coenzyme Q but are relatively deficient in phosphatidylglycerol, diphosphatidylglycerol and neutral lipid (mainly diacylglycerol) compared with the cytoplasmic membrane; in addition, the internal membrane fatty acids are more saturated, but the functional significance of these changes is not known.

3B.1.6 Storage vesicles

Many plants accumulate energy stores in the form of triacylglycerol (Secs 3E.2.1 & 6B.1), which may comprise as much as 95% of the dry weight of lipid-rich seeds. The majority of the fat bodies are surrounded by a monolayer ('half-unit') membrane, which has been isolated in some instances. These fat bodies, which are the only organelles to possess a high content of non-membranous lipid, have enormous economic importance. For example, in the USA soya beans are used to produce about 8000 tonnes of oil per annum, while the total value of soya bean exports represent approximately 5% of the total export income!

The membrane of plant oil bodies appears to be derived from the endoplasmic reticulum. During oil accumulation by seeds, the increased lipid content of the tissue is primarily accounted for by a large increase in the number of new oil bodies. Several workers have suggested that the term spherosome is synonymous with oil body. However, it is now reasonably certain that the plant spherosome, although rich in lipid (mainly phospholipid), does not have a role in energy storage but, instead, acts as a source of the vacuoles which develop into storage grains.

Although the internal organs of plants contain very little wax, certain rare plants such as jojoba accumulate large amounts of wax ester as an energy reserve. These wax bodies have a rather similar morphological appearance to the oil bodies of other plants. Waxes and hydrocarbons are widespread but usually minor constituents of freshwater and some marine algae. Although wax esters play an important role in the marine food chain, the initial organisms in this chain, the phytoplankton algae, do not make wax esters but merely serve as food for herbivorous copepods which do make them and are the principal food of fish.

In fungi, triacylglycerols are the major constituents of oil droplets suspended in the mycelial or spore cytoplasm, but their relative abundance depends on the culture conditions and stage of growth. Triacylglycerols are often by far the most predominant lipid in fungi. For example 92% of the total lipid of some *Tricholoma* spp. and 80% of some *Candida* spp. can be triacylglycerol.

Unlike plant (and animal) cells, bacteria do not contain acyl lipid stores, and it is believed that their acyl lipids are confined to membranes, although it should be said that this has rarely been rigorously demonstrated. Lipid droplets are infrequently seen in bacteria. An exception, that could have ecological and economic importance, are the hydrocarbon inclusions present in hydrocarbon-oxidising bacteria. These inclusions in *Acinetobacter* spp. grown on hexadecane consist of a droplet of lipid (largely neutral lipid: 50% wax ester, 18% alcohol, 14% glycerides, 12% alkane and 6% fatty acid) surrounded by a membrane that appears as a single track in the electron microscope. The inclusions contain phospholipids (mainly phosphatidylethanolamine and phosphatidylglycerol), which may constitute the membrane in the form of a phospholipid monolayer with the acyl chains dipping into the neutral lipid core. Interestingly, the inner and outer membranes of the cell envelope of this Gram-negative organism also contain significant amounts of alcohols and wax esters, which are particularly common in *Acinetobacter* spp.

Generally they appear to be membrane components, although it has been suggested that, as in marine algae, they may act as energy stores.

The most usual form in which lipid is stored in bacteria is as granules of poly-β-hydroxybutyrate (PHB) (Sec. 6B.2), which are found in a wide range of species, both Gram-negative and Gram-positive. Granules of PHB may constitute a large proportion of the cellular dry weight under certain conditions, e.g. up to 70% in *Azotobacter beijerinckii*! The granules are surrounded by a 'half-unit membrane' that appears as a single track in the electron microscope; however, it may not be a phospholipid membrane in view of the very small amounts of phospholipid in PHB granules.

3B.2   Acyl composition of lipids

3B.2.1   Introduction
The fatty acyl composition of membrane lipids can vary at least as much as the proportions of the parent lipids. By a combination of chemical and enzymic analyses it is now possible to determine the positional distribution and combinations of fatty acids at the various ester linkages in any given lipid – provided that sufficient material can be isolated and the experimenter has enough patience! Fortunately, for many purposes such detail is unnecessary.

In general terms the lipids of algae, yeast and higher plants are rich in polyunsaturated fatty acids. This is particularly true of marine species. In contrast, bacteria, with very few exceptions, do not contain polyunsaturated fatty acids, although monounsaturated fatty acids are commonly present.

3B.2.2   Plants
In plant membranes the major saturated fatty acid is invariably palmitic, and there are usually significant quantities of stearic, while oleic acid is the major monoenoic acid, and linoleic almost the only significant dienoic acid. The membranes of chloroplasts contain exceptionally high levels of α-linolenic acid – up to 90% of the total fatty acids in some chloroplast lamellae! In addition, the other membranes of leaves contain higher amounts of linolenate when compared to equivalent fractions isolated from the shoots or roots of the same plant. One subclass of plants is termed the '16 : 3-plants' because they contain significant amounts of hexadecatrienoic acid. An example of such a species is a widely used experimental tissue, spinach leaves.

It is when individual lipids are examined that really striking fatty acid patterns emerge. For example, the storage triacylglycerols of seeds are the location of various bizarre fatty acids found in nature. Some examples are given in Table 3.9. By contrast, the fatty acid contents of individual lipids in leaf tissue are rather consistent (Table 3.9). In chloroplast lamellae the two galactosylglycerides both contain exceptionally high levels of α-linolenate (Table 3.10). When hexadecatrienoic acid is present it is found only in diacylgalactosylglycerol and not in diacyldigalactosylglycerol in spite of the fact that the latter is derived biosynthetically from diacylgalactosylglycerol. Linolenic acid is also a major component of the two negatively-charged lamellar lipids, diacyl-sulphoquinovosylglycerol and phosphatidylglycerol. However, in both of these lipids palmitate is also an important constituent. Studies with lipases, which

Table 3.9 A comparison of the fatty acid compositions of plant leaves and storage tissues.

	Fatty acid composition (%, w/w)					
	16:0	18:0	18:1	18:2	18:3	Other
storage tissue						
soya bean	11	4	25	52	8	trace
rape seed	4	1	13	7	7	59[a]
castor bean	1	—	4	5	—	90[b]
green leaves						
all varieties	12–16	1–4	2–7	8–24	48–66	3–6

[a] Mainly erucic acid (cis-Δ13-docosenoic acid).
[b] Mainly ricinoleic acid (Δ12-hydroxy oleic acid).

hydrolyse fatty acids specifically from either the 1- or the 2- position of the glycerol moiety of diacylsulphoquinovosylglycerol or phosphatidylglycerol, indicate that the majority of the palmitate is attached to the 2-position whereas linolenate is at the 1-position. This distribution, with the unsaturated acid on the 1-position, contrasts with the situation usually found in animal and bacterial lipids (i.e. 1-saturated, 2-unsaturated). In addition, the unusual fatty acid, *trans*-Δ3-hexadecenoic, is localised exclusively at the 2-position of phosphatidylglycerol. Extra-chloroplastic higher plant membranes, with their predominant phospholipid constitutents, also contain large amounts of polyunsaturated fatty acids, although this is less exaggerated than in the chloroplast. Examples for the major phospholipid, phosphatidylcholine, are included in Table 3.11, and compared with the phosphatidylcholine of some algae and bacteria. Phosphatidylethanolamine in plants has a rather similar fatty acid composition to phosphatidylcholine, while phosphatidylinositol seems to be characterised by rather high levels of palmitate.

3B.2.3 Algae
It is also of interest that the cyanobacteria with their 'chloroplast-like' lipid composition have their fatty acid positional distributions determined by chain length rather than unsaturation. Thus, phosphatidylglycerol and diacyl-

Table 3.10 The fatty acid compositions of chloroplast lamellar lipids.

Plant	Lipid[a]	Fatty acid composition (%, w/w)						
		16:0	16:1[b]	16:3	18:0	18:1	18:2	18:3
spinach	MGDG	trace	—	25	—	1	2	72
	DGDG	3	—	5	—	2	2	87
	SQDG	39	trace	—	trace	1	7	53
	PG	11	32	—	trace	2	4	47
barley	MGDG	4	1	—	trace	1	3	90
	DGDG	9	2	—	1	3	7	78
	SQDG	32	3	—	1	2	5	55
	PG	18	27	—	3	2	11	38

[a] For abbreviations see Table 3.4.
[b] Palmitoleic acid in glycolipids and *trans*-Δ3-hexadecenoic acid in PG.

Table 3.11 The fatty acid composition of phosphatidylcholine from different sources.

Source	Fatty acid composition (%, w/w)					
	16:0	16:1	18:0	18:1	18:2	18:3
spinach chloroplasts	12	—	—	9	16	58
pea seed microsomes	4	1	4	46	43	2
castor bean mitochondria	17	—	8	18	53	5
Chlorella vulgaris cells	16	3	5	9	44	18
Fucus serratus cells[a]	15	—	—	14	16	10
Paracoccus denitrificans cells	8	1	4	86	—	—
Rhodospirillum rubrum chromatophores	34	18	7	39	—	—
Rhodopseudomonas sphaeroides chromatophores	7	2	3	77	—	—

[a] Rich in very-long-chain polyunsaturated fatty acids, e.g. 20:4 and 20:5.

sulphoquinovosylglycerol contain 16C fatty acids at the 2-position and 18C acids at the 1-position. As in chloroplasts, this results in an enrichment in unsaturation at the 1-position, because the main 18C acids are oleate, linoleate, and γ-linolenate or α-linolenate depending on algal species, whereas the main 16C acid is palmitate. Although the cyanobacteria carry out light-induced CO_2 fixation and oxygen evolution, their phosphatidylglycerol does not contain *trans*-Δ3-hexadecenoate, unlike higher plants and green algae (Table 3.12). It will be seen in Table 3.12 that the fatty acid compositions of the major lipids of *Chlorella* are very similar to those of higher plant chloroplast lamellae (Table 3.10). Although brown algae contain many non-chloroplastic membranes, their lipid composition is dominated by the three glycosylglycerides which contain significant quantities of the very-long-chain polyunsaturated fatty acids (Table 3.12). The relatively high concentration of palmitate in the phosphatidylglycerol and

Table 3.12 The fatty acid compositions of algal lipids.

Alga (species)	Lipid	Fatty acid composition (%, w/w)							
		16:0	16:1	18:1	18:2	18:3	18:4	20:4	20:5
green	MGDG	5	1	3	17	45	—	—	—
(*Chlorella vulgaris*	DGDG	8	3	3	35	37	—	—	—
grown at 25 °C)	SQDG	32	5	10	25	15	—	—	—
	PG	31	21	10	25	5	—	—	—
	PC	16	3	9	44	18	—	—	—
cyanobacteria	MGDG	23	22	4	12	33	—	—	—
(*Anabaena variabilis*	DGDG	18	23	2	9	39	—	—	—
grown at 22 °C)	SQDG	53	3	10	12	20	—	—	—
	PG	51	2	4	13	30	—	—	—
brown (*Fucus serratus*	MGDG	6	1	6	4	16	25	9	22
grown at 10 °C)	DGDG	12	2	9	8	11	11	11	16
	SQDG	40	1	24	5	9	trace	3	6

Only the major lipids are shown. 16:1 is *trans*-Δ3-hexadecenoic acid for PG from *Chlorella vulgaris* and palmitoleic acid (*cis*-Δ9-hexadecenoic acid) for other lipids. 18:4 contains double bonds in the Δ6,9,12,15 positions, 20:4 is mainly arachidonic acid and 20:5 is mainly the Δ5,8,11,14,17 isomer. For abbreviations see Table 3.4.

diacylsulphoquinovosylglycerol of higher plant chloroplasts, when compared to the fatty acid composition of the galactosylglycerides, can also be seen clearly in the algae. In addition, marine algae often contain significant quantities of myristate; this usually represents 5–15% of the total fatty acids, rising to 30% in some diatoms.

3B.2.4 Yeasts

The formation of functional mitochondria in yeasts referred to above (Sec. 3B.1.3) has been studied in relation to the fatty acid content of the phospholipids. Of the three major fatty acids present in the promitochondria of *Saccharomyces cerevisiae*, palmitate showed no change (20% of the total acids), palmitoleate a huge increase (from 7% to 44%) and oleate a corresponding decrease (from 62% to 34%) upon differentiation of mitochondria.

3B.2.5 Bacteria

As discussed above, there is generally very little difference in the phospholipid composition of the membrane types in a bacterial cell, and likewise the fatty acyl composition of the major phospholipids is usually the same. For instance, there are few differences in acyl composition of the major phospholipids of chromatophores and cytoplasmic membranes of purple bacteria, the latter being slightly more unsaturated. In *Rhodopseudomonas capsulata* and *Rhodopseudomonas sphaeroides* the major fatty acid is *cis*-vaccenic $(18:1\ (11c))$ (80–85% of the total), and its levels are the same in photosynthetic and non-photosynthetic membranes (see Table 3.8). The diacylsulphoquinovosylglycerol of purple bacteria is noteworthy in this respect because it has a much more saturated fatty acid $(16:0 + 18:0)$ composition than the major phospholipids (Table 3.8).

It is important to note that the acyl composition of bacterial lipids is normally very dependent upon the cultural conditions (see Sec. 3E) so that when there are fatty acyl changes, such as temperature-dependent changes in unsaturation index, they generally occur in all classes of phospholipids. Cultural variations are probably the major cause of discrepancies in different analyses of the same organism (see Ch. 6).

By comparison the fatty acid composition of the lipopolysaccharide of Gram-negative bacteria is quite distinct from that of the phospholipids (Table 3.13). It contains large amounts of hydroxy fatty acids (mostly 2- and 3-hydroxymyristate) together with laurate, myristate and some palmitate; in contrast, hydroxy acids are usually absent from the phospholipids which contain mainly 16C and 18C saturated and monounsaturated fatty acids. Thus hydroxy acids are used as 'markers' for lipopolysaccharide. However, they do occur in the ornithine lipids of those relatively few bacteria that contain them (Sec. 2B.5); and in some gliding bacteria up to 50% of the phospholipid acyl chains are hydroxy branched-chain fatty acids, which may help to produce a flexible membrane for their particular method of movement.

Differences are also seen when the phospholipids of the inner and outer membranes are compared (Table 3.13). The phosphatidylethanolamine of the outer membrane is more saturated than that of the inner membrane, which is reflected in its higher viscosity as measured using electron spin resonance spectroscopy. In *Escherichia coli* the outer membrane contains more $1–16:0$, $2–16:1$-phosphatidylethanolamine and less $1–18:1,2–16:1$-phosphatidyl-

Table 3.13 A comparison of the fatty acid compositions of phospholipids and lipid A of *Escherichia coli.*

Fatty acid	Fatty acid composition (%, w/w)			
	Total phospholipid	Inner membrane phospholipid	Outer membrane phospholipid	Lipid A
12:0	—	—	—	9
14:0	3	3	4	10
16:0	35	34	37	2
16:1	33	33	31	1
17:0 cyc	2	2	3	—
18:0	1	1	1	—
18:1	25	26	23	1
19:0 cyc	1	1	1	—
D-3-OH-14:0	—	—	—	77

ethanolamine than does the inner membrane. But the distribution of label from radioactive fatty acids incorporated into the various molecular species of phosphatidylethanolamine is the same for the two membranes. (The phosphatidylethanolamine is synthesised in the inner membrane and translocated to the outer membrane.) It is assumed, therefore, that the differences arise as a result of differential turnover rates for the molecular species. A phospholipase A_1 is present in the outer membrane, but it is not known if this is the enzyme responsible.

The major outer membrane protein in Gram-negatives is the so-called 'murein lipoprotein'. This protein has a cysteine residue at its N-terminus that is amide-linked to predominately palmitate, and thioester-linked to a diacylglycerol. The acyl groups on the diacylglycerol are essentially the same as those of the phospholipids.

The fatty acyl components of lipopolysaccharide and murein lipoprotein (and lipoteichoic acid and lipomannan) in bacteria are collectively known as 'bound lipid', because they are not freely extractable with organic solvent mixtures.

3B.2.6   Fungi

The cell walls of fungi contain appreciable quantities of lipids (up to 20% in some *Aspergillus* spp.) with bound lipids generally being present in two to three times greater abundance than the freely extractable lipids. It has been suggested that they have a structural role in the cell wall. Triacylglycerols, sterols and sterol esters are usually major components and phospholipids sometimes so. The predominant acyl constituents are palmitoleate and oleate.

3C   Intramembrane lipid distribution

3C.1   Introduction

There is now considerable evidence that biological membranes display sidedness not only with regard to their protein components, but also as regards their lipid constituents (see Fig. 3.4 for two examples). Lipid asymmetry in membrane

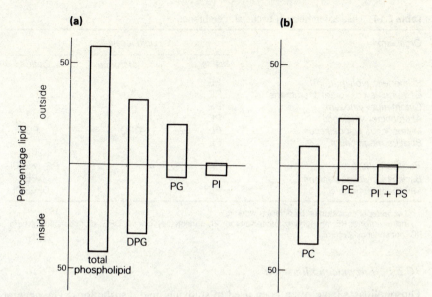

Figure 3.4 Asymmetry of membrane lipids: (a) plasma membrane (*Micrococcus lysodeikticus*) and (b) glyoxysomal membrane (castor bean). DPG = diphosphatidylglycerol, PG = phosphatidylglycerol, PI = phosphatidylinositol, PC = phosphatidylcholine, PE = phosphatidylethanolamine, PS = phosphatidylserine.

bilayers raises important questions regarding the specificity of the insertion of newly synthesised lipids, and must depend on either a very low frequency of exchange from one surface to the other or a controlled reciprocal exchange. The specific arrangements of lipids within membranes has obvious functional implications.

3C.2   Methods used to determine lipid asymmetry

The sided-distribution of lipids in the membrane bilayer has been probed in five main ways. These are:

(a)   the enzymic modification of lipids
(b)   the chemical modification of lipids
(c)   exchange of lipids
(d)   physicochemical techniques
(e)   immunochemical techniques.

In the case of the plant and bacterial cytoplasmic membranes, it should be remembered that, unlike animal cells, conversion to protoplasts (or spheroplasts) is necessary before any digestion or labelling experiments can be undertaken. It has to be assumed that removal of the cell wall does not induce changes in the lipid distribution across the cytoplasmic membrane. On the other hand, chloroplast membranes can be isolated and studied directly.

Table 3.14   Lipid asymmetry in bacterial membranes.

Organism	Lipid localisation		
	Inside	No preference	Outide
Salmonella typhimurium	PE		
Escherichia coli (outer membrane)	PE		
Chromatium vinosum	PE		
Azotobacter vinelandii	PE		
Micrococcus lysodeikticus	PI	DPG	PG
Bacillus megaterium[a]	PE		
			PE
Bacillus subtilis			PE, lyso-PG
Bacillus amyloliquefaciens			PE
Mycobacterium phlei	PE		DPG, PI

[a]Two independent studies by different workers.
Abbreviations: PE, phosphatidylethanolamine; PI, phosphatidylinositol; DPG, diphosphatidylglycerol; PG, phosphatidylglycerol.

3C.2.1   Enzymic modification

Phospholipases have often been used in studying lipid distributions. The general idea is to digest lipids in the outer layer only and quantify the lyso-derivatives produced. But the technique suffers from a number of problems, such as destabilisation of the membrane structure, a marked substrate specificity of the phospholipases and the induction of trans-bilayer lipid movement. The data for *Bacillus megaterium* and other bacilli (Table 3.14) illustrate the problems of this approach (and that of chemical labelling – see below) in that different results can be obtained, often when the *same* bacterium is investigated by different workers.

However, certain enzymes can sometimes be used – e.g. lactoperoxidase-mediated iodination in *Acholeplasma laidlawii* revealed that the glycosylglycerides were in the outer half of the bilayer whereas phosphatidylglycerol, diphosphatidylglycerol and phosphoglycolipids were equally distributed.

3C.2.2   Chemical modification

Non-enzymic modification of lipids can be carried out sometimes. For instance, the primary amino group of phosphatidylethanolamine may be modified by compounds such as trinitrobenzenesulphonic acid. This reagent will modify only lipids in the outer membrane half provided that the membrane is *impermeable* to the compound. This may not always be true and careful controls have to be carried out. Just as with enzyme treatments, there may be differences in the reactivity of different lipid molecules – e.g. saturated molecular species are slower to react. The chemical modifiers may also perturb the membrane structure and this may prevent reactions going to completion. Chemical labelling has shown the aminoacylphosphatidylglycerol is largely on the inner face of the cytoplasmic membrane of *Acholeplasma*.

3C.2.3   Lipid exchange

The purification of a number of water-soluble lipid-exchange proteins enables the location of specific phospholipids in the outer half of a membrane to be tested. The lipid to be measured is specifically exchanged with the same lipid from liposomes and, provided that one of the membrane systems has been radioactively labelled, then the amount of the phospholipid can be calculated. Naturally, the exchange proteins have to be absolutely specific with regard to substrate and the rate of transbilayer movement of lipids has to be low.

3C.2.4 Physicochemical methods

The most useful technique here is that of nuclear magnetic resonance (usually phosphorus-31 nuclear magnetic resonance). For technical reasons, sonicated vesicles have to be used and paramagnetic ions are used as chemical shift or broadening agents to modify signals from those lipids in contact with the ions. For example, if the shift agent is added to a vesicle suspension it will modify preferentially the signals from lipids in the outer half of the bilayer.

3C.2.5 Immunochemical techniques

These provide only qualitative information and are limited by the few phospholipids which are antigenic. For example, antibodies have been used to locate phosphatidylglycerol in mycoplasma membranes, and also lipopoly-saccharide in the outer membrane of Gram-negative bacteria. The outer membrane of Gram-negative bacteria represents a particularly asymmetric membrane in that only the outer half of the membrane bilayer contains the lipopolysaccharide, whose polysaccharide chains face the exterior where they function as, for instance, bacteriophage or bacteriocin receptors.

Antibodies have also been used to examine the lipids of plant chloroplasts. The negatively charged phosphatidylglycerol and diacylsulphoquinovosylglycerol were found to be concentrated in the outer surface while the neutral galactosylglycerides were concentrated in the inner.

An approach that has not been fully exploited with bacteria as far as lipids are concerned, is the use of rightside-out and inside-out membrane vesicles. Rightside-out vesicles can be made by gently lysing osmotically-sensitive protoplasts or spheroplasts, whilst inside-out vesicles result from disrupting bacteria with the French pressure cell.

3D Subcellular fractionation and membrane isolation

3D.1 Plants

An initial difficulty in the preparation of pure and undamaged organelles from higher plants is in releasing the cell contents. The cellulose walls of plant tissues are so tough that shear forces sufficient to break them may also rupture fragile organelles. In addition, many hydrolytic enzymes or chemicals such as phenolics and acids present in the cell may be liberated during homogenisation. Protective agents may be added to the homogenisation medium to minimise any harmful effects of these substances. For example, polyvinylpyrrolidone is used to complex polyphenols and quinones, and diethylpyrocarbamate will inhibit RNAse.

In order to separate organelles with reasonable purity, density gradient centrifugation usually has to be employed. Although a large number of possible materials can be used, a few have found particular favour. Sucrose is commonly used but is osmotically active and this can lead to damage of organelles. In many situations sorbitol, which is not usually metabolised and is less harmful, is better. Ficoll (a low density sucrose polymer) may be used for iso-osmotic gradients and is particularly useful for delicate organelles, like chloroplasts from *Euglena*. Recently, silica sols have been used and permit the very rapid isopycnic separation of labile organelles such as chloroplasts. With this technique intact chloroplasts are separated from those which have a damaged outer envelope.

Mention has already been made of the presence of many hydrolytic enzymes in plant tissues. Lipolytic acyl hydrolases, neutral lipases, phospho- or glycolipid lipases and lipoxygenase (see Ch. 5) are of particular interest so far as lipids are concerned. Many of these enzymes are not only present in large amounts but are also particularly active against the endogenous lipids found in higher plant organelles. In order to minimise the deleterious effects of these lipases, precautions have to be taken. Not least amongst these is the choice of tissue. For example cabbage leaves (high in phospholipase D) or potato tubers (normally rich in lipolytic acyl hydrolase, but see Table 3.6) are poor tissues to use for the preparation of organelles, whereas spinach leaves (which are low in lipolytic enzyme activity) can yield chloroplasts which show little evidence of lipid degradation, at least over a short period. Other precautions which can be taken include working at a slightly alkaline pH and the addition of chelating agents, antioxidants or bovine serum albumin.

Recently, enzyme mixtures have been prepared which will digest the cell walls of plants, bacteria and algae. The resulting protoplasts can be isolated and then gently ruptured in order to yield undamaged intracellular organelles or particles. Unfortunately, the cell walls of plants and algae are often rather resistant to commercially-available cellulase or other enzymes, so this promising method is not yet widely used.

Not only will endogenous lipases hydrolyse lipids, they will, of course, damage organelles. A commonly used extraction system for lipids is the so-called Folch procedure. This method involves homogenisation of the tissue in a mixture of chloroform and methanol. When the technique is applied to plant tissues, it frequently yields phosphatidic acid, unesterified fatty acids and fatty acid peroxidation products – none of which are normally present in any more than trace amounts. The galactosylglycerides are particularly vulnerable because of the substrate specificity of the lipolytic acyl hydrolase and because of their high content of polyenoic fatty acids. Furthermore, even at 50–60 °C the activity of phospholipase D in methanolic solutions can give rise to a new lipid – phosphatidylmethanol! To avoid these problems, plant tissues should be treated either with steam or boiling isopropanol in order to inactivate the causative enzymes.

3D.2 Bacteria and other micro-organisms

Like plant cells, bacteria are also surrounded by a tough, rigid cell wall, which must either be broken or removed before cellular membranes can be isolated. Peptidoglycan, the rigid polymer of the cell wall of most bacteria, can be digested using lysozyme (for staphylococci the enzyme lysostaphin has to be used instead). If an osomotic stabiliser such as sucrose is present, Gram-positive bacteria are converted to protoplasts. In Gram-negative bacteria, the peptidoglycan is protected by the outer membrane which can be disrupted using ethylenediaminetetraacetate (EDTA); the combined action of lysozyme and EDTA produces an osmotically-sensitive spheroplast (this differs from a protoplast in having remnants of the outer membrane still attached to the cell). Protoplasts and spheroplasts are the starting point for the isolation of intracellular membrane or the separation of the inner and outer membranes of the Gram-negative cell envelope.

In Gram-positive bacteria, particularly if they are plasmolysed prior to lysozyme treatment, mesosomes are extruded from the cells during conversion to protoplasts. The mesosomes detach from the plasma membrane at an appropriate Mg^{2+} concentration (which is species-specific). They can then be collected by differential centrifugation or purified directly by centrifugation in sucrose, Ficoll or CsCl gradients.

The inner and outer membranes of Gram-negative bacteria are structurally and functionally quite distinct, and it is of considerable interest, therefore, to be able to separate and isolate them in a pure form. Although broken cells or cell envelope preparations have been used, the usual starting material is lysed spheroplasts (see above). As long as spheroplast formation is complete, the inner and outer membranes can be separated by sucrose isopycnic density gradient centrifugation (the outer membrane banding at a higher density than the inner membrane).

Photosynthetic bacteria are widely used to study many aspects of photosynthesis. There are many published procedures for isolating chromatophores from purple bacteria, but most consist essentially of disrupting cells and then separating the chromatophores from larger cell wall and cytoplasmic membrane fragments by differential centrifugation. This gives a crude chromatophore suspension, purifiable by density gradient centrifugation, when chromatophores band at a higher density than the other cellular membranes. Sometimes, gel filtration chromatography on Sepharose columns or rate-zonal sedimentation procedures have been used. It is important that all buffers contain <10 mM Mg^{2+} because divalent cations promote the aggregation of chromatophores with each other and the cell membrane. In the past, the use of buffers containing Mg^{2+} has led to the artefactual production of 'light' and 'heavy' chromatophore fractions, the latter being due to the association of true chromatophores with envelope membrane fragments.

Similar methods have been employed to isolate chlorosomes from green sulphur bacteria (Sec. 3B.1.4), but they have been less well studied and characterised. In addition, membrane fractions can be obtained from unicellular algae and fungi. Generally, the cell wall is digested enzymically before gentle cell breakage and differential centrifugation. Particular interest has centred on the isolation of photosynthetic lamellae from algae and of (pro)mitochondria from yeast.

3E Factors affecting the lipid composition of plants and micro-organisms

3E.1 Introduction

Many changes in growth conditions, as well as the age of a given organism, will affect its lipid content and metabolism. There are good reasons for many of these alterations and these can be directly related to particular functions of lipids in cells. This functional aspect of lipid composition is dealt with more fully in Chapter 6, and we will devote ourselves here to an outline of the sorts of factors which can change lipid composition in a given organism.

3E.2 Development

3E.2.1 Plants

During the life-cycle of a plant cell, changes will occur in its morphology due to the appearance and disappearance of membranes, organelles and storage products. Lipids are clearly involved as major membrane constituents and may also be present as storage material.

As plant seeds germinate, lipid synthesis is 'switched on' when the water content rises due to imbibition. Not all the enzymes of lipid synthesis are active immediately and so the first types of, for example, fatty acids may not accurately reflect later synthetic patterns. Components of the outer cuticle have to be made first, followed by the constituents of photosynthetic membranes. As the shoot elongates and the first leaves appear, so more of the total lipid synthesis of the plant is concentrated on the formation of glycosylglycerides and polyunsaturated fatty acids – the major constituents of chloroplast membranes (Table 3.15).

Many seeds contain stored carbohydrates and these reserves are mobilised to yield acetyl-CoA and glycerol 3-phosphate precursors for lipid formation. As soon as active chloroplasts are developed then these will take over to supply photosynthate for lipid synthesis. Some seeds (e.g. soya bean and sunflower) contain stores of lipids which is almost always triacylglycerol. This lipid is broken down to yield carbohydrate by β-oxidation and the glyoxylate pathway. This process requires the transient formation of glyoxysomes during the first week of germination (see Secs 5B.2 and 6B.1).

Table 3.15 Summary of changes in lipid content and metabolism during the life of a plant.

Period	Lipid composition and changes
germination	seed composition reflects stored lipid; may be 90% of seed dry weight in oil-rich seeds; cuticular lipids and waxes produced first, then leaf lipids, especially those for chloroplast membranes (MGDG, DGDG, SQDG, polyunsaturated fatty acids, *trans*-Δ3-hexadecenoic acid, chlorophylls, carotenoids); oil-rich seeds show disappearance of triacylglycerol reserves as glyoxysomes are present briefly
mature plant	steady-state metabolism takes place with typical leaf, shoot and root compositions maintained
flowering	many specialised pigments, e.g. carotenoids, produced for colouring petals
seed development	in oil-rich seeds, lipid accumulation in three phases
senescence	rapid degradation of chloroplast acyl lipids and pigments

Abbreviations: MGDG, diacylgalactosylglycerol; DGDG, diacyldigalactosylglycerol; SQDG, diacyl-sulphoquinovosylglycerol.

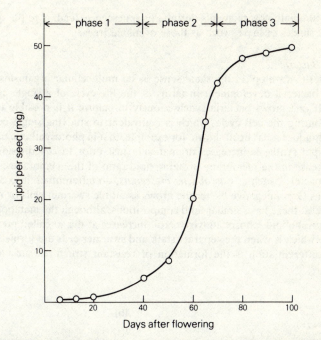

Figure 3.5   Changes in the lipid content of developing soya bean seeds.

In leaf tissue some 70% of the total protein and 80% of the total lipid are present in chloroplasts. Any changes in chloroplast membranes, therefore, will usually be reflected by equivalent alterations to leaf total lipids. As leaves mature their lipid composition becomes constant, although rapid rates of metabolism indicate that this pattern requires constant maintenance (Table 3.15).

Flowering and the setting of fruit can bring about large changes in lipid metabolism. If the developing seed accumulates lipid then the sudden and temporary appearance of many enzymes involved in lipid synthesis is a notable feature of this developmental period. Characteristically, the development of a seed occurs in three phases (Fig. 3.5). The first period, depending upon the species and environmental conditions, lasts 10–30 days and is characterised by a very slow accumulation of lipids. During this period the seed has a high moisture content. The second period, lasting 2–5 weeks, is characterised by the rapid accumulation of lipids. In the third stage, ending at seed maturity, only minor amounts of lipid generally accumulate but there is a gradual loss of moisture. If the seed contains unusual fatty acids in its stored lipid (e.g. ricinoleate in castor bean) then these acids are first made during the second phase of seed development.

Senescence in leaves is accompanied by very rapid losses in those lipids which are components of chloroplasts. Thus, the glycosylglycerides are quickly deacylated and the liberated linolenic acid is particularly susceptible to lipoxygenase-catalysed attack. The damage to the chloroplast also allows oxidation of the photosynthetic pigments to take place with the consequent yellowing of leaves (Table 3.15). Some plants accumulate new pigments during this senescence, leading to the spectacular autumnal shades of leaves such as those of maple trees.

3E.2.2 Bacteria

Few bacteria 'develop' in the same sense as do multicellular organisms. One can consider bacterial development in terms of the life-cycle of a single cell or of a culture. If one grows bacteria synchronously in culture it is possible to monitor changes during the cell cycle, which is equivalent to studying single cells. Using these techniques it has been shown, for example, that in photosynthetic bacteria the rate of lipid synthesis increases dramatically just prior to cell division giving a 60% decrease in the membrane protein : lipid ratio of the membranes.

Some bacteria, such as *Caulobacter crescentus*, do differentiate into distinct cell types. This Gram-negative bacterium grows as motile swarmer cells or non-motile stalked cells; these have similar lipid compositions, although the metabolism of the major phospholipid, phosphatidylglycerol, increases at the so-called pre-divisional cell stage which is when the separate stalk and swarmer cells are formed. Another type of differentiation is the formation of resistant structures such as cysts or

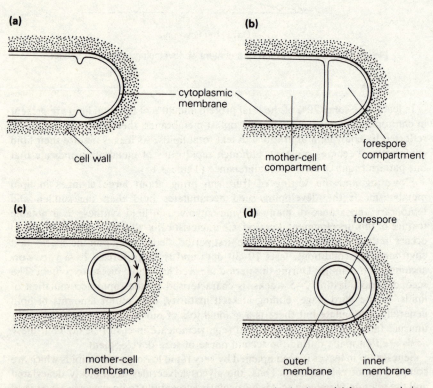

Figure 3.6 Formation of the double membrane system in the bacterial forespore during sporulation. (a) Septum formation initiated; (b) separation of forespore and mother-cell; (c) engulfment of forespore; (d) formation of double membrane system.

spores. When *Azotobacter vinelandii* encysts large amounts of resorcinols and pyrones are accumulated, accounting for approximately 16% of the cyst dry weight.

A particularly intriguing example of membrane differentiation in bacteria occurs in bacilli and corynebacteria that produce endospores. Sporulation involves the formation of a forespore which differentiates into the mature spore. Forespores possess two membranes, both derived from the cytoplasmic membrane of the mother-cell (Fig. 3.6). These membranes have opposite orientations, the outer forespore membrane having an orientation that is opposite to both the inner forespore membrane and the mother-cell cytoplasmic membrane due to the mode of formation of the double membrane system (Fig. 3.6). The forespore membranes have a low lipid : protein ratio, particularly the outer membrane, and both are enriched in diphosphatidylglycerol (~35% of the total lipid) compared with the mother-cell cytoplasmic membrane (~4%). The spore membranes also differ in their lipid composition: the inner membrane contains most of the spore phospholipid (~80%), while the outer membrane contains more of the neutral lipid (~60%), and in the inner membrane there is a red carotenoid not found in other spore or cell membranes. The mobility of the phospholipid in spores is considerably reduced, possibly by interaction with Ca^{2+}, and phospholipid fluidity increases during spore germination.

When bacteria are grown in batch culture (rather than in continuous culture) the growth conditions change, particularly as the end of logarithmic growth is reached and the stationary phase is entered. These changes are often accompanied by alterations in the amount and/or the type of lipids, particularly those reflecting an 'end-point' of metabolism – e.g. an increase in diphosphatidylglycerol (at the expense of phosphatidylglycerol) or cyclopropane fatty acids (at the expense of unsaturated fatty acids). It has been argued that it is preferable to grow bacteria in continuous culture, but even though this produces a relatively constant environment there are often changes in lipid composition dependent upon factors such as the limiting nutrient used and the dilution rate.

3E.2.3 Fungi

Fungal growth and development can be divided into three stages – spore germination, vegetative growth and reproduction.

Spore germination is stimulated by a number of low molecular weight lipids, the nature of the stimulating compounds varying with the fungal species. For example, nonanal stimulates *Puccinia graminia* and isocaproic acid stimulates *Agaricus bisporus*. Many spores contain lipid bodies or globules. During germination these bodies tend to be reduced in size or disappear but become prominent in sporogenous hyphae and developing spores. During germination, the stored lipid of the granules is oxidised even if, as with stored phospholipid, it is a rather poor substrate for oxidation. Not only does stored lipid provide energy by oxidation, it may also participate in the glyoxylate cycle (see Sec. 5B.2), by which fat is converted to carbohydrates.

During the vegetative growth phase, fungi are influenced by various environmental factors including temperature, inorganic nutrients, oxygen and pH. In general the lipid content of the mycelium increases rapidly during the vegetative growth of most species. This is due mainly to accumulation of triacylglycerols and phosphoglycerides. In many species the content of sterols also increases to a 'threshold' which it has been suggested is required for sporulation – a process accompanied by loss of sterols. There has been much interest in the role of steroids in fungal reproduction. This stemmed from the observation that there was a sterol requirement for oospore and zoosporangium formation in non-steroid producing fungi such as *Pythium* spp. In addition, the steroid antheridiol acts as a hormone involved in the sexual reproduction of *Achlya bisexualis*.

3E.3   Effects of growth conditions on lipid composition

3E.3.1   Light

The effects of light can be distinguished from those of other growth parameters, particularly in plants where light is absolutely essential for mature growth and for the synthesis of particular lipids. In contrast, the photosynthetic bacteria do not rely upon light conditions because they are able to grow heterotrophically in the dark. Nevertheless, as in plants and other photosynthetic organisms, light produces spectacular morphological effects as the specialised photosynthetic membranes are developed.

In plants there is a general increase in glycosylglycerides and phosphatidylglycerol, the lipids that characterise chloroplastic membranes. Although α-linolenic acid is synthesised in the dark, its concentration increases significantly on light-exposure. The same is true of carotenoids. However, the most characteristic changes are the appearance of *trans*-Δ3-hexadecenoic acid (esterified in phosphatidylglycerol) and the chlorophylls (Table 3.16). The rates of various reactions in lipid synthesis (such as those of fatty acid synthetase) are also increased by light exposure but some of these effects are due to the increased supply of cofactors.

Because algae contain chloroplasts, their lipid changes in the presence of light are similar to those of plants (although a few green algae can synthesise chlorophylls in the dark). But in photosynthetic bacteria, although there are large increases in total phospholipid and pigment contents during the switch from heterotrophic to phototrophic growth, the compositional changes are less; the main difference is the increased content of phosphatidylglycerol in photosynthetic membranes (see Sec. 3B.1.4). The green photosynthetic bacteria differ from the purple, because they have specific photosynthetic organelles called chlorosomes (Fig. 3.2) that contain diacylgalactosylglycerol, which is not found in other cell membranes, thus giving photosynthetically-grown cultures a very different lipid composition (Table 3.16).

3E.3.2   Temperature

The effect of growth temperature has been studied in a large number of different biological systems. In plants, algae and fungi lower growth temperatures generally lead to an increase in fatty acid unsaturation. In bacteria, yeasts and cyanobacteria unsaturation changes also occur, often in combination with chain length changes; in bacteria there may be alterations in the amount and type of branching, alone or in combination with other changes (Table 3.17). Studies with marine *Vibrio* spp. show that even closely related organisms may react quite differently to shifts in environmental temperature, changing either chain length or unsaturation. There may also be small temperature-dependent adjustments in the total amount of membrane phospholipid.

Table 3.16 Light-induced changes in the lipids of photosynthetic organisms.

Organism	Change	Lipid alterations
plants	greening of etiolated tissue	increase in chloroplast acyl lipids; appearance of chlorophyll, 16:1(3t)
protozoa, e.g. *Euglena gracilis*	heterotrophic → phototrophic growth (chloroplast development)	appearance of chlorophyll; increase in proportions of chloroplast lipids
photosynthetic bacteria	heterotrophic → phototrophic growth	increase in pigments and proportions of lipids of photosynthetic membranes
(purple)	(internal membrane development)	
(green)	(chlorosome development)	appearance of diacylgalactosyl-glycerol

As a rough generalisation psychrophilic members of a genus have a higher proportion of shorter and/or monounsaturated fatty acids than do mesophilic members, which have longer and/or fewer unsaturated fatty acids.

The often quoted temperature-dependent change in fatty acid unsaturation in *Escherichia coli* in fact conceals a chain length change, because it is the synthesis of 18 : 1Δ11 (but not 16 : 1Δ9) that is regulated (see Sec. 4A.2). In *Pseudomonas fluorescens* and *Salmonella typhimurium* (Table 3.17) the opposite is true – i.e. 16C fatty acids rather than 18C change. As well as these changes in fatty acid biosynthesis, there are probably effects of temperature on the acyltransferases, so that the final phospholipid acyl composition is modified by growth temperature in two ways.

All of these temperature-dependent changes in acyl composition have been rationalised in terms of membrane fluidity and its maintenance (Sec. 6A.2). However, this may not always be the correct explanation. For example, careful studies in some plant tissues (e.g. seeds) have shown that the main effect of lowering temperature is to increase oxygen solubility and, hence, the amount available for desaturation. This is patently not true of bacteria that use the

Table 3.17 Effects of growth temperature on some bacterial fatty acid compositions.

Fatty acid	Percentage composition					
	Salmonella typhimurium		*Escherichia coli*		*Micrococcus cryophilus*	
	17 °C	37 °C	20 °C	40 °C	0 °C	20 °C
12:0	15	19				
14:0	19	16	2	3		
15:0	6	6				
15:0 br	19	20				
16:0	14	22	21	37	1	1
16:1	15	8	13	7	53	21
17:0 cyc			8	23		
18:0	2	1			3	2
18:1	8	7	49	15	53	76
19:0 cyc			7	15		

anaerobic pathway of unsaturated fatty acid biosynthesis. Changes in oxygen availability could explain the effects of temperature on many bacteria which contain oxygen-dependent desaturases, because in very few instances have rigorous experiments using continuous cultures (where the oxygen tension can be controlled) been performed. However, despite this caveat, when this approach has been used it has been demonstrated that the environmental factor that is 'sensed' is temperature rather than oxygen tension.

Changes in fatty acyl unsaturation may occur directly *in situ* in the membrane, because phospholipids can act as desaturase substrates (Sec. 4A.3). In contrast, acyl chain length changes can only occur after deacylation of phospholipids, since the fatty acids are elongated at the carboxyl terminus. Because the new fatty acid frequently has to be made *de novo*, chain length changes are usually slower than changes in unsaturation. Changes in branching also appear to be 'slow' and are mediated by new acyl chain synthesis during growth rather than turnover. Some bacilli demonstrate both types of response to temperature. For example, *Bacillus megaterium* grown at 35 °C contains no unsaturated fatty acids, but on shifting to 20 °C a Δ5 desaturase is induced; such large temperature shifts may produce a hyperinduction state, resulting in the rapid synthesis of unsaturated fatty acids, probably by desaturation of membrane lipids. This transient change (the desaturase is also thermolabile) is followed by slower but more permanent changes in the chain length and type of branching of the major odd-chain length fatty acids dependent upon 'addition synthesis' involving membrane growth and cell division. In this way these bacteria are able to cope with short- and long-term fluctuations in temperature.

3E.3.3 Salt concentration and water activity

In plants both salt stress and exposure to drought lead to an overall increase in all root lipids due to a proliferation of internal membranes. A similar change is seen on brief cold-exposure of some species which are 'frost resistant'. In addition, membranes from such plants usually have a higher lipid : protein ratio than those from non-stressed plants.

As far as tolerance to salt is concerned, there are essentially four kinds of bacteria. The first ('non-halophiles') comprises most bacteria which can tolerate low salt concentrations (usually up to approximately 0.5 M). The second kind of bacterium (halotolerant) can grow in up to 3.0 M salt. The third consists of 'moderate halophiles' (e.g. *Vibrio costicola*) which need some salt to grow and can withstand up to approximately 3.5 M salt. The fourth group are the 'extreme halophiles' (mainly *Halobacterium* and *Halococcus* spp.) which require high salt concentrations (>4 M) for growth, and which contain unusual phytanyl ether lipids (Sec. 2F).

It is the first three groups, which contain 'normal' lipids – e.g. phosphatidylethanolamine and phosphatidylglycerol are the major phospholipids – that show marked effects of salt concentration on lipid composition. Non-halophiles such as staphylococci generally respond to increasing salt concentration by synthesising more diphosphatidylglycerol at the expense of phosphatidylglycerol. In moderate halophiles, e.g. *Vibrio costicola*, the proportion of phosphatidylglycerol increases relative to phosphatidylethanolamine, probably in order to maintain the charge balance of the membrane in the face of the high

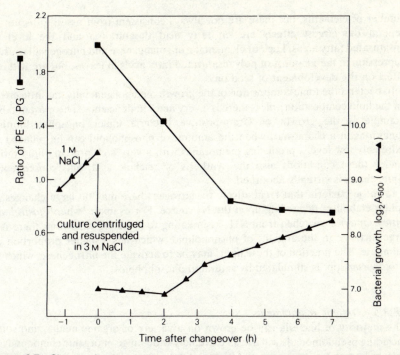

Figure 3.7 Changes in phospholipid composition in the moderately halophilic bacterium *Vibrio costicola* after an increase in salt concentration. A culture of *V. costicola* growing exponentially in medium containing 1 M NaCl was centrifuged and the bacterial pellet resuspended gently in fresh medium containing 3 M NaCl. Control experiments show that there is no effect on phospholipid composition and growth rate due to either centrifugation or resuspension.

external Na^+ concentration. It appears that such changes are necessary for growth, because they occur during the adaptation period before growth recommences after a shift in salt concentration (Fig. 3.7), and this change probably requires new enzyme synthesis.

When salt concentration is varied the water activity also is changed. Just as for temperature and oxygen solubility (see above), it is not always clear which parameter is the one responsible for initiating the lipid compositional changes. More rigorous experiments are required to resolve this question.

Water activity can also be lowered by high concentrations of non-ionic solutes such as glycerol or sugar; these are most commonly found in preserved foods. Organisms which inhabit these environments are called osmotolerant, and usually are yeasts or fungi rather than bacteria. There is no evidence that they have specific membrane lipids: for example, the lipids of the osmophilic *Saccharomyces rouxii* are not markedly different from other *Saccharomyces* spp.

3E.3.4 Specific inorganic requirements

In plants, deficiencies of nitrogen and minerals will lead to a decline in nutritional status, metabolism and, hence lipid synthesis. Such effects have been noted for a number of elements, but these are not always consistent from tissue to tissue. In general the largest effects are on fatty acid desaturation and the level of unsaturated fatty acids. Lack of elemental iron, manganese and nitrogen all cause a depression in the amounts of polyunsaturated fatty acids in leaves, but are without effect on the development of seed oils.

In bacteria the ionic composition of the growth medium generally has little effect on the lipid composition unless there is a very marked deficiency. The most notable example is the growth of Gram-positive bacteria under phosphate-limiting conditions in a chemostat, when the amount of phospholipid can be reduced to extremely low levels, producing membranes with a very high protein : lipid ratio. Under these conditions also the synthesis of teichoic acid, and consequently lipoteichoic, is virtually abolished.

In those bacteria that facultatively fix nitrogen there may be large changes in lipid metabolism depending upon the N_2 source. For example, when *Clostridium pasteurianum* is switched from NH_3-assimilating to N_2-fixing conditions there is a large increase in the amount of phospholipid, which has a higher proportion of palmitate. The function of these lipids may be to activate the nitrogenase, which in *Azotobacter* spp. is stimulated by saturated phospholipids.

3E.3.5 Organic requirements

The majority of bacteria can be grown on a variety of organic media, and some including pseudomonads will use an extremely wide range of organic compounds as sole carbon and energy source. Although lipid composition may vary with carbon source, there is no consistent pattern from one nutrient or organism to another. In addition, there are often greater differences induced by other cultural parameters, including the stage of growth in batch culture, so that it is extremely difficult to compare results from different studies.

There are, however, some clearer relationships between lipid composition and medium composition. In Gram-positive bacteria that contain branched-chain fatty acids their proportion (relative to straight-chain fatty acids) and the relative amounts of *iso-* and *anteiso*-branched fatty acids, can be varied by the addition of valine, leucine or isoleucine to the growth medium. These amino acids serve as precursors for odd-numbered straight-chain, *iso-* and *anteiso*-branched fatty acids respectively (see Sec. 4A.2).

Bacteria will often take up lipids, e.g. sterols, fatty acids or some detergents, from the growth medium and incorporate them, unchanged or after modification, into their cell membranes. For example, staphylococci grown in serum-containing media contain small amounts of sterols, which has led to several erroneous reports of these lipids as normal (i.e. synthesised by bacteria) membrane constituents. The mycoplasmas require both sterols and fatty acids in the medium to support their growth; they are unable to synthesise either. The same is true also of some *Treponema* spp. This requirement has led to the widespread use of mycoplasmas as model systems for the study of the effects of membrane lipid composition on membrane function, because their fatty acid and sterol composition can be varied between such wide limits.

Growth of bacteria on alkanes can modify dramatically the lipid composition – e.g. growth on hexadecane leads to large rises in the proportion of palmitate in lipids. In some species of *Acinetobacter* intracellular membrane-bound lipid inclusions are formed (Sec. 3B.1.6).

3E.3.6 pH

Micro-organisms are present in environments of an extremely wide range of pH – i.e. from zero to >11 – and include representatives of yeasts, fungi, algae, bacteria and cyanobacteria. Acid environments are more common than alkaline ones, and often they are hot, e.g. the interior of coal tips or sulphur springs – so the inhabitants are thermophilic as well! Acidophiles contain large amounts of cyclised fatty acids (Sec. 4A.5) and amino phospholipids, which may help to exclude H^+, but it is unclear whether these are an adaptation to the acidic environment. Similarly, the phytanyl ether lipids present in archaebacterial acido- and alkaliphiles, and extreme halophiles (Sec. 2F), may be the result of their evolution from a distinct ancestor rather than adaptation *per se*.

Some organisms that grow at neutral pH alter their lipid composition in response to less drastic changes in pH. For example, during the growth of *Staphylococcus aureus* the pH falls due to acid production and there is a rise in the amount of aminoacylphosphatidylglycerol (Sec. 2B.2), but the function of this lipid is not proven.

3E.4 Effects of herbicides, antibiotics and other chemicals on lipid composition

Several substituted pyridazinone herbicides cause severe inhibition of lipid synthesis in plants. Where the compounds block carotenoid formation, their use causes bleaching in treated tissues because chlorophyll is photo-oxidised in the absence of protecting carotenes. Some of these herbicides have also been found to prevent α-linolenate formation.

Another group of herbicides which inhibit lipid formation are certain thiocarbamates (e.g. diallate and triallate). These act by preventing very-long-chain fatty acid synthesis and, hence, the supply of precursors for wax formation. Plants treated in this way are killed because excessive transpiration takes place through the defective cuticular layer. Trichloroacetic acid has a similar effect.

Quite the other side of the picture is presented when we consider herbicides and bacteria. Although certain compounds do inhibit bacterial growth or are even bactericidal, it is soil bacteria (and other micro-organisms) that are generally responsible for the degradation of excess herbicide from direct spraying or decaying vegetation. The persistence of a herbicide is related to its solubility and to the ease with which it is degraded by the soil microflora. Between them, the members of this flora are capable of degrading an extremely wide range of chemical structures, including aromatic rings and substituents such as halides.

There are no widely used antibiotics that specifically inhibit bacterial lipid metabolism, which is surprising, perhaps, in view of the differences between bacteria and animals in this respect. Bacitracin, which is used in some topical ointments, blocks the conversion of undecaprenol diphosphate to the mono-phosphate, and so should more properly be regarded as an inhibitor of cell wall and other extracellular polysaccharide synthesis. Diazoborine inhibits

CMP–KDO transferase in lipopolysaccharide biosynthesis (Sec. 4D.3), but is too toxic to be used therapeutically.

The polymixins, which have a limited clinical use, interact with cell membranes and alter their permeability probably by disorganising the lipid bilayer. They are more active against Gram-negatives, possibly because of their higher content of phosphatidylethanolamine.

3E.5   Genetic control of lipid composition

The genetic control of lipid synthesis is particularly important in the area of seed oils. Breeding trials have included those for flax, rape, safflower and sunflower. These have been particularly successful in the case of rape seed where the normal 'wild' type accumulates large amounts (45% of the total fatty acids) as erucate. Erucic acid (22 : 1, 13*c*) can produce toxic effects in animals and is considered undesirable in large amounts for human consumption; varieties of rape have been bred which accumulate only 2–5% of this acid.

One of the reasons why such few species of bacteria have been studied so intensively is because of the accumulated knowledge of the genetics of organisms such as *Escherichia coli* and *Pseudomonas aeruginosa*. As in other areas of biochemistry the use of mutants has helped considerably in defining metabolic pathways of lipids as well as their functions, and examples have been cited where appropriate; these include mutants with defective genes of fatty acid degradation (Sec. 5B.2) or biosynthesis (Sec. 4A.2). Recently several genes of phospholipid biosynthesis (Sec. 4B.2.2) have been cloned, making possible further genetic analysis and studies in regulation.

4

Biosynthesis

4A Fatty acids

4A.1 Acetyl-CoA carboxylase

Irrespective of the nature of their individual fatty acid synthetases, all plants and micro-organisms that synthesise fatty acids *de novo* require the enzyme acetyl-CoA carboxylase and a source of acetyl-CoA in order to provide the necessary malonyl-CoA.

Although they fulfil similar metabolic functions and carry out essentially the same chemical reactions (viz. the ATP-, Mg^{2+}- and biotin-dependent carboxylation of acetyl-CoA), the acetyl-CoA carboxylases of animals and bacteria differ considerably in their structure and regulation (Table 4.1). The acetyl-CoA carboxylase from animal tissues exists as a multifunctional inactive protein (MW 240 000), which consists of three domains – biotin carboxylase, biotin carboxyl carrier protein (BCCP) and carboxyltransferase – in a single polypeptide chain. In the presence of citrate this protomer rapidly aggregates to form polymers of MW $7 \times 10^6 - 10 \times 10^6$, which are the active form of the enzyme.

In contrast, the acetyl-CoA carboxylase of *Escherichia coli* (the only bacterial species to have been studied in detail) readily dissociates into its constituent proteins, biotin carboxylase, biotin carboxyl carrier protein (BCCP) and carboxyltransferase, all of which have been purified to homogeneity and retain full activity. Biotin carboxylase and BCCP are both dimers of identical subunits, whilst carboxyltransferase is a tetramer of two different subunits (A_2B_2). They carry out the following reactions:

$$\text{BCCP–biotin} + \text{HCO}_3^- + \text{ATP} \underset{\substack{\text{biotin} \\ \text{carboxylase}}}{\overset{Mg^{2+}}{\rightleftharpoons}} \text{BCCP–biotin–CO}_2^- + \text{ADP} + \text{P}_i$$

$$\text{BCCP–biotin–CO}_2^- + \text{acetyl–CoA} \underset{\substack{\text{carboxyl-} \\ \text{transferase}}}{\rightleftharpoons} \text{BCCP–biotin} + \text{malonyl–CoA}$$

Inside the bacterial cell it is assumed that the three proteins form a multienzyme complex (that would be structurally equivalent to the animal protomer but functionally equivalent to the polymer).

Besides this structural and organisational difference, the bacterial enzyme is also controlled differently to the animal enzyme, which is activated by a citrate-induced

Table 4.1 Comparison of acetyl-CoA carboxylases from different sources.

Property	Animal	Bacterial	Plant	
			Chloroplast	Seed
structure	polymeric filaments (MW = 7×10^6– 10×10^6) of protomers (MW $\simeq 2 \times 10^5$) of a single multifunctional protein	multienzyme complex (MW = 2.7×10^5) of three proteins	multienzyme complex of three proteins	complex (MW = 3.5×10^5) of two proteins
intermolecular interactions	protomers are resistant to dissociation	readily dissociable	readily dissociable	carboxyltransferase can be isolated from complex, but the carboxylase and BCCP[a] are not separable
regulation	activated by citrate and inhibited by palmitoyl-CoA	inhibited by (p)ppGpp	inhibited by heat-stable factor, ADP/ATP	regulated by Mg^+ and K^+

[a] BCCP = biotin carboxyl carrier protein.

polymerisation of the protomers and is also regulated by phosphorylation of the enzyme protein. Citrate has no effect on bacterial acetyl-CoA carboxylase and there is no phosphorylation control. The bacterial enzyme is regulated instead by the two nucleosides guanosine 3'- diphosphate, 5'-diphosphate (ppGpp) and guanosine 3'-diphosphate, 5'-triphosphate (pppGpp); these reduce enzyme activity by inhibiting carboxyltransferase. The guanosine nucleosides, which are unique to bacteria, are formed by phosphoryl transfer from ATP to GDP or GTP on the ribosome in response to amino acid starvation, or when other conditions of reduced growth rate lead to ribosome 'idling'. Their main effects are on RNA and protein biosynthesis, and they have a central role in regulating growth rate and cell division. They regulate lipid metabolism by inhibiting not only acetyl-CoA carboxylase but probably phospholipid biosynthesis also (see below). To control lipid metabolism at the level of acetyl-CoA carboxylase makes good 'energetic sense', because its product, malonyl-CoA, is used almost exclusively for long-chain fatty acid biosynthesis.

The reason for different regulatory controls in animals and bacteria becomes clearer when the major fates of fatty acids are considered. In animals a large proportion are used to synthesise triacylglycerols, which are an energy store, and the source of cytoplasmic acetyl-CoA for fatty acid synthesis is citrate from mitochondria, the primary site of energy production; thus citrate acts as a positive feed-forward effector. In bacteria, the main source of acetyl-CoA is from pyruvate (through the action of pyruvate dehydrogenase), and most fatty acids are used in the biosynthesis of membrane phospholipids; therefore, regulation of fatty acid synthesis is through (p)ppGpp, whose levels reflect overall cell growth, including membrane biogenesis.

Compared with bacteria and animals, the origin of acetyl-CoA for the carboxylase enzyme in plants remains somewhat controversial, even though isolated reports of pyruvate dehydrogenase in plastids have been made. Carefully purified chloroplasts contain a very active acetyl-CoA synthetase, and indeed it is the presence of this enzyme that allows biochemists to use radioactive acetate in order to study fatty acid synthesis! In the plant, acetyl-CoA synthetase may be used to esterify acetate which has been generated from O-acetylserine. Thus, the initial precursor of fatty acids in plants may be derived from several sources apart from the direct action of pyruvate dehydrogenase.

In most plants acetate is a rare anion. Moreover, acetate is relatively inert and so can be removed from the site of its synthesis, the mitochondrion, to the chloroplasts without being lost to other metabolic pathways. Once in the chloroplast it is rapidly activated by the acetyl-CoA synthetase in the stroma and can then be carboxylated.

Acetyl-CoA carboxylase enzymes in higher plants represent an intermediate level of organisation compared with the polymeric organisation of animals and the easily dissociated bacterial proteins. In wheat germ the biotin carboxyl-carrier protein and the carboxylase are associated with one fraction while the carboxyltransferase can be isolated independently. Although the enzyme shows aggregation patterns similar to those of animal enzymes, citrate is not involved and there is no change in specific activity upon aggregation. Instead, small changes in Mg^{2+} and K^+ concentrations can profoundly affect the activity and these may occur either as the result of ion transport or chelation (e.g. by ATP) during metabolism. The acetyl-CoA carboxylase from barley embryo has similar properties but the enzyme in chloroplasts has 'prokaryotic' properties in that the individual proteins of the complex can be isolated separately in an active form (Table 4.1). All three constituent proteins of the acetyl-CoA carboxylase are soluble and located in the chloroplast stroma. It has recently been found that AMP and ADP are competitive inhibitors, with respect to ATP, of plant acetyl-CoA carboxylase. Because ATPase and adenylate kinase are also present, the levels of adenylate nucleotides may alter markedly the activity of acetyl-CoA carboxylase *in vivo*. This has led to an explanation as to why fatty acid synthesis in isolated chloroplasts typically is stimulated 20-fold by light (Fig. 4.1). Thus, during active photosynthesis, the amounts of available substrates (especially NADPH) increase and, in addition, the activity of acetyl-CoA carboxylase is increased because ADP levels drop as ATP is generated. Apart from the possible controlling effects of ions on carboxyl-transferase activity, which have already been referred to, quite a number of photosynthetic plant tissues contain an unidentified heat-stable, high molecular weight inhibitor of the carboxyltransferase enzyme. This compound may function as an *in vivo* regulator of activity in much the same way as citrate and acyl-CoAs are implicated in the control of mammalian acetyl-CoA carboxylases.

The acetyl-CoA carboxylase of yeast again appears to be an intermediate form. It has a high molecular weight (~600 000) and is probably composed of four similar or identical subunits. It can be prepared in a form that is activated by citrate, but this activation is not accompanied by polymerisation of the enzyme.

Euglena gracilis, grown under photoautotrophic conditions, contains a multienzyme complex composed of acetyl-CoA carboxylase, phosphoenolpyruvate kinase and malate dehydrogenase in the ratio 1 : 25 : 500. The malate

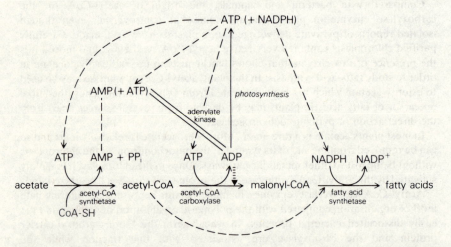

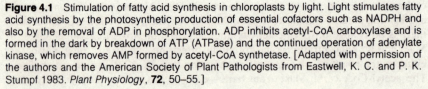

Figure 4.1 Stimulation of fatty acid synthesis in chloroplasts by light. Light stimulates fatty acid synthesis by the photosynthetic production of essential cofactors such as NADPH and also by the removal of ADP in phosphorylation. ADP inhibits acetyl-CoA carboxylase and is formed in the dark by breakdown of ATP (ATPase) and the continued operation of adenylate kinase, which removes AMP formed by acetyl-CoA synthetase. [Adapted with permission of the authors and the American Society of Plant Pathologists from Eastwell, K. C. and P. K. Stumpf 1983. *Plant Physiology*, **72**, 50–55.]

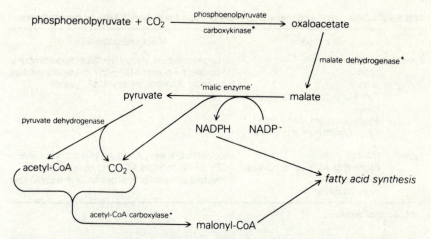

Figure 4.2 Metabolic function of the acetyl-CoA carboxylase complex of *Euglena gracilis*. The three enzymes asterisked are enzymes of the complex.

dehydrogenase is an isoenzyme unique to the multienzyme complex, which is believed to co-operate *in vivo* with the so-called 'malic enzyme'; this converts malate to pyruvate and CO_2, forming NADPH in the process. The CO_2 is used by acetyl-CoA carboxylase to form malonyl-CoA, the substrate (with NADPH) for fatty acid synthesis (Fig. 4.2). The purpose of the multienzyme complex may be to regulate in some way the activity of acetyl-CoA carboxylase by providing the CO_2 in the correct amount and at the appropriate site inside the cell.

4A.2 Fatty acid synthetase

Classic experimental work in the 1950s and 1960s on fatty acid synthesis in *Escherichia coli* and yeast clearly defined the individual steps by which two-carbon fragments are added from malonyl-CoA to a primer molecule until a long-chain fatty acid is produced. The chemical reactions appear to be essentially the same in all synthetases, consisting of successive cycles of condensation of activated acceptor with malonyl-CoA, reduction, dehydration and a second reduction until the required chain length is required. The fatty acid is then released as a free acid (animals) or as an acylthioester (yeast, plants and bacteria). The mechanistic details of these reactions are dealt with in standard biochemistry textbooks: we shall concern ourselves largely with comparative aspects.

Fatty acid synthetases have been categorised into Type I, Type II and Type III synthetases; the distribution and some of the properties of the Type I and Type II enzymes are summarised in Table 4.2. Although there is no strict phylogenetic pattern, the Type I synthetases tend to occur in higher organisms and the Type II synthetases in lower organisms. The Type I synthetases are typically found in animals, but are also present in various microbes, including yeast and fungi and the two bacterial genera, *Mycobacterium* and *Corynebacterium*. In this respect *Euglena gracilis* is particularly interesting, because when it is grown heterotrophically (when it is 'animal-like') it contains a Type I synthetase; when it is grown

Table 4.2 Distribution and major properties of Type I and Type II fatty acid synthetases.

	Distribution	Major characteristics
Type I	animals birds yeast *Euglena gracilis* (cytoplasm) fungi *Mycobacterium smegmatis* *Corynebacterium diphtheriae*	(a) large molecular weight, multifunctional proteins; (b) covalently-bound ACP;[a] (c) release unesterified fatty acids (animals) or acyl-CoA (yeast)
Type II	plants *Euglena gracilis* (chloroplasts) most bacteria cyanobacteria	(a) dissociable enzymes (six) in a complex with ACP; (b) dissociable ACP; (c) release acyl- thioesters or transfer final products directly to lipid

[a] Acyl carrier protein.

photoautotrophically it contains, in addition to its cytoplasmic Type I enzyme, a Type II synthetase in the chloroplasts. Higher plant synthetases have been purified from several photosynthetic tissues and appear to be Type II enzymes, very similar to those present in most bacteria. Whereas Type I and Type II synthetases are responsible for the *de novo* synthesis of a long-chain fatty acid (usually palmitate) from a primer which is normally acetyl-CoA, the elongation of preformed acyl chains is carried out by Type III synthetases; these usually use malonyl-CoA as the source of the 2C units, and are often termed 'elongases' to distinguish them from *de novo* synthetases. Interestingly, in contrast to the Type I and Type II systems, many of the Type III synthetases are membrane bound.

The main feature that distinguishes the Type I and Type II synthetases is the functional organisation of their constituent proteins. Type I synthetases contain large molecular weight multifunctional polypeptide chains – i.e. they possess several active sites of different enzymic specificity – and the acyl carrier protein (ACP) is covalently bound to the protein. On the other hand, Type II synthetases consist of individual enzymes that can be isolated in an active form and the ACP readily dissociates from the synthetase. It is assumed, however, that inside the cell the individual enzymes of the synthetase associate to form a (loosely bound?) multienzyme complex. The site of association may be the cell membrane because, in *Escherichia coli* at least, ACP is localised in the membrane's vicinity.

Type I synthetases are characteristically high molecular weight proteins $(0.4 \times 10^6 - 2.5 \times 10^6)$ comprising two or more large multifunctional polypeptide chains (MW $1.8 \times 10^5 - 2.7 \times 10^5$). In animals it is generally accepted that the two chains are identical. Recent genetic analysis of fatty acid synthesis mutants of yeast has demonstrated that there are two unlinked polycistronic genes, designated *fas* 1 and *fas* 2. The *fas* 1 gene codes for the acetyltransferase, malonyl (palmitoyl) transacylase, dehydrase and enoyl reductase enzymes, while *fas* 2 codes for the phosphopantetheine-binding region and the β-ketoacyl synthetase and reductase enzymes. The Type I synthetase of yeast is much larger (2.4×10^6) than that of animals $(0.4 \times 10^6 - 0.5 \times 10^6)$, whilst that of *Euglena gracilis* (1.7×10^6) is intermediate. That of *Corynebacterium diphtheriae* (2.5×10^6) is particularly large. The significance of these differences is unknown and much remains to be

Table 4.3 A comparison of Type I fatty acid synthetases.

Property	Animal	Yeast	Mycobacteria
molecular weight	$0.4 \times 10^6 - 0.5 \times 10^6$	2.4×10^6	2.0×10^6
subunit arrangement	A_2	A_6B_6	A_{6-8}
malonyl (palmitoyl) transacylase	separate catalytic sites	same catalytic site	same catalytic site
end product	unesterified fatty acid	acyl-CoA	bimodal distribution of 16C and 24C fatty acids

discovered of the intramolecular and intermolecular organisation of these complex multifunctional proteins. The yeast synthetase is probably an A_6B_6 complex, made up of two different multifunctional proteins (A and B), although only five ACP-binding sites have been identified. Animal synthetases are presumed to be A_2 complexes in view of their molecular weights. These and other differences between the yeast and animal synthetases are summarised in Table 4.3.

The Type I fatty acid synthetase of *Mycobacterium smegmatis* (Table 4.3) is unusual in several respects: the two reductases have different reduced pyridine nucleotide specificities, β-ketoacyl-ACP reductase requiring NADPH and enoyl-ACP reductase requiring NADH (other Type I enzymes use NADPH only); the products have a bimodal distribution with peaks at 16C and 24C, and the overall rate of fatty acid synthesis is increased by two types of poly-methylsaccharides found in the complex mycobacterial wall. The main effect of the polymethylsaccharides is to increase the proportion of 16C products. It is not known how they function, but they are present in large amounts in the mycobacterial cell and can form stoichiometric complexes with acyl-CoAs of chain lengths 14–20C by hydrophobic interaction between the acyl chain and the helical methylated polysaccharide chain. This may well affect both elongation of the acyl chain and its release from the synthetase. It is believed that the bimodal pro-duct pattern arises from the overlapping specifications of the condensing enzyme (broad specificity) and the transacylase (specificity for 16C acyl chains). High acetyl-CoA : malonyl-CoA ratios also increase the proportion of 16C products, probably by favouring the 'priming reaction' with acetyl-CoA relative to elongation with malonyl-CoA. The synthetase is also regulated by palmitoyl-CoA, which inhibits the enzyme by causing it to dissociate, and by the polymethylsaccharides, which activate by promoting aggregation. The Type I enzymes of animals and yeast are also inhibited by palmitoyl-CoA. The Type I complex of *Aspergillus fumigatus* is unusual in that it is markedly inhibited by malonyl-CoA, but it is not known if this is a general property of fungal synthetases.

Most of what we know about Type II synthetases comes from studies on *Escherichia coli*, not only because the seven proteins of the complex have been purified but also because mutants are available. Indeed the *Escherichia coli* enzyme is used as a mechanistic model for fatty acid synthetases in general, which is perhaps unfortunate because it possesses certain unique features, not the least among them being the fact that it synthesises both saturated and unsaturated fatty acids. The reason for this is the presence of a β-hydroxydecanoyl-ACP β, γ-dehydrase, which produces *cis*-3-decenoyl-ACP (the precursor of unsaturated fatty acids), instead of *trans*-2-decenoyl-ACP (the precursor of saturated fatty acids). This β, γ-dehydrase can be specifically inhibited and its role in forming unsaturated fatty acids was confirmed by the isolation of *fab* A mutants, which lack a functional enzyme and consequently require unsaturated fatty acids for growth.

The presence of these two dehydrase enzymes forms a branch-point at the 10C level (Fig. 4.3). The unsaturated fatty acids are produced because the double bond in *cis*-3-decenoyl-ACP is not reduced as it has the wrong configuration for enoyl-ACP reductase; this enzyme can reduce a *trans*-2-decenoyl-ACP substrate, which therefore leads to a saturated fatty acid. The β-hydroxydecanoyl-ACP dehydrase is essentially specific for a 10C substrate so that the major unsaturated fatty acids in *Escherichia coli* are 16 : 1Δ9 cis (palmitoleic) and 18 : 1Δ11 *cis* (*cis*-vaccenic), with only small amounts of 16 : 1Δ7 *cis* and 18 : 1Δ9 *cis* which are derived from a 12C intermediate. The positions of these double bonds are quite characteristic and are even diagnostic of the so-called '**anaerobic pathway**', compared with the unsaturated isomers that are produced by aerobic desaturation (Sec. 4A.3).

The *Escherichia coli* fatty acid synthetase is also unusual in that it contains two β-ketoacyl-ACP synthetase enzymes (known as synthetase I and II – not to be confused with Type I and II fatty acid synthetases!). Both enzymes can be used for the condensation reactions of saturated fatty acid synthesis; the β-ketoacyl-ACP synthetase I is used in the elongation steps from *cis*-decenoyl-ACP leading to palmitoleic acid, whereas β-ketoacyl-ACP synthetase II is relatively specific for the

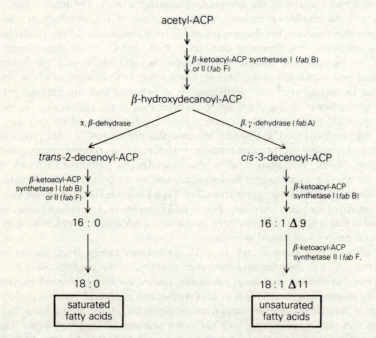

Figure 4.3 The anaerobic pathway of fatty acid biosynthesis in bacteria, showing mutants of *Escherichia coli*. ACP = acyl carrier protein.

elongation of palmitoleic to *cis*-vaccenic acid (Fig. 4.3). The synthetase II is important in thermal regulation of membrane phospholipid acyl composition, because this enzyme is more temperature-sensitive than the synthetase I. This fact, together with its substrate specificity, results in a relative increase in synthetase II activity at lower growth temperatures. Thus the proportion of unsaturated fatty acids increases as the flow of substrate down the 'unsaturated arm' of the anaerobic pathway (Fig. 4.3) rises relative to that down the 'saturated arm'.

The presence of two β-ketoacyl-ACP synthetase enzymes went unnoticed until the discovery of an *Escherichia coli* mutant, *fab* B, that was shown to contain synthetase II, but not synthetase I, which is coded for by the *fab* B gene. Another class of mutants (*fab* F) do not increase their *cis*-vaccenate content in response to a decrease in growth temperature, and lack synthetase II. Thus it was possible to show clearly that there were two enzymes coded by separate genes. The enzymes have now been distinguished on the basis of their different molecular weights, amino acid compositions and peptide maps, pH optima and heat sensitivity as well as their different substrate specificities; the synthetase I is extremely sensitive to the antibiotic cerulenin, while the synthetase II is relatively resistant.

The fatty acid synthetase of *Brevibacterium ammoniagenes* is particularly interesting because it combines properties of Type I and Type II fatty acid synthetase enzymes. It is a high molecular weight (1.2×10^6) complex, like Type I synthetases, but, like the *Escherichia coli* Type II synthetase, it produces both saturated and unsaturated fatty acids. The mechanism appears to be similar to that of *Escherichia coli* except that the branch point is probably at 12C (not 10C as in *Escherichia coli*), because the major product is $18 : 1\Delta9$ *cis* instead of $18 : 1\Delta11$ *cis*.

The major fatty acids in most Gram-positive and some Gram-negative genera of bacteria are branched-chain, *iso* or *anteiso* fatty acids (see Sec. 3A.1.1). The molecular nature of the synthetase that makes branched-chain fatty acids and the mechanism of the elongation reactions using malonyl-CoA and NADPH are typical of Type II enzymes. The major difference lies in the preference of branched-chain fatty acid synthetases for a short-chain, branched acyl-CoA (instead of acetyl- or butyryl-CoA) as the primer. This distinction arises from the specificity of the acyl-CoA : ACP transacylase enzymes. The significance of this is obvious when it is realised that it is the nature of the primer that determines the type of fatty acid product (see Fig. 4.4).

The transaminases which convert the amino acids to their respective keto acids (Fig. 4.4) are probably specific for a particular amino acid. However, the same enzyme, α-ketoisovalerate dehydrogenase, converts all three ketoacids to their respective CoA-thioesters, which enter the elongation cycle after conversion to ACP-thioesters by the acyl-CoA : ACP transacylase. In some *Bacillus* spp. the relative affinites of this enzyme for the various CoA-thioesters determine the ratio of the branched products. It may also convert small amounts of acetyl- or butyryl-CoA to their ACP-thioesters, which accounts for the small amounts of straight-chain 14C and 16C fatty acids in such organisms. In other respects the control of branched fatty acid synthesis is poorly understood. It is closely linked to branched amino acid metabolism, and the relative amounts of different branched fatty acids can be varied in certain bacilli by changing the availability of these amino acids.

A major fatty acid in *Mycobacterium phlei* is 10-methyl stearic acid, which is

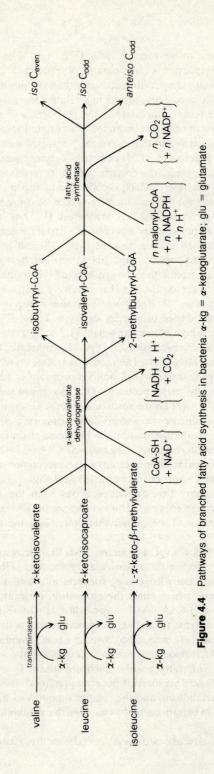

Figure 4.4 Pathways of branched fatty acid synthesis in bacteria. α-kg = α–ketoglutarate; glu = glutamate.

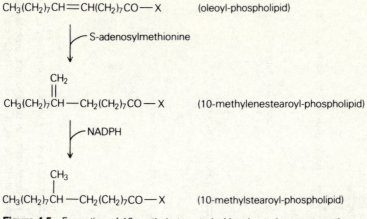

$CH_3(CH_2)_7CH=CH(CH_2)_7CO—X$ (oleoyl-phospholipid)

S-adenosylmethionine

$\underset{\|}{CH_2}$
$CH_3(CH_2)_7CH—CH_2(CH_2)_7CO—X$ (10-methylenestearoyl-phospholipid)

NADPH

$\underset{\|}{CH_3}$
$CH_3(CH_2)_7CH—CH_2(CH_2)_7CO—X$ (10-methylstearoyl-phospholipid)

Figure 4.5 Formation of 10-methyl stearate in *Mycobacterium smegmatis*.

formed, not by *de novo* synthesis, but by methylation of oleic acid (probably as a phospholipid acyl chain) with *S*-adenosylmethionine followed by reduction with NADPH (Fig. 4.5). This type of reaction is similar to that involved in the synthesis of cyclopropane fatty acids (see below). The biosynthesis of other methylated fatty acids (in mycobacteria) is dealt with in Section 4D.1.2.

In higher plants, the product of their Type II fatty acid synthetase is palmitoyl-ACP. Unlike the *Escherichia coli* enzyme the plant synthetase does not form unsaturated fatty acids. The constituent enzymes have been purified recently from several sources. The two reductases from spinach leaves differ in their cofactor requirements. The β-ketoacyl-ACP reductase preferentially uses NADPH while the enoyl-ACP reductase has an absolute requirement for NADH (cf. the *Mycobacterium smegmatis* Type I fatty acid synthetase). In addition, as in *Escherichia coli*, two β-ketoacyl-ACP synthetase enzymes have been purified. One will condense a broad range of primer units (up to 14C) and is sensitive to cerulenin, whereas the second is specific for palmitoyl-ACP and is inhibited by arsenite. The latter enzyme can, therefore, be regarded as the key part of the Type III synthetase, palmitate elongase. In addition to being elongated, palmitoyl-ACP produced by the higher plant fatty acid synthetase may be hydrolysed by a thioesterase, transferred to an acyl lipid or desaturated. Since stearate is not a very prevalent fatty acid in plant tissues, it is obvious that the stearoyl-ACP formed by palmitate elongase is usually either desaturated or elongated. Elongation of stearate (probably as stearoyl-CoA) leads to the formation of the very-long-chain saturated fatty acids. In contrast to palmitate elongation which is sensitive to arsenite, stearate elongation is inhibited by fluoride and thiocarbamate herbicides. The very-long-chain fatty acids are components of cuticular wax and are precursors of most of the other cutin constituents (see Sec. 4D.2). They are also involved in the formation of the other important protective covering in plants – the suberin of root or wounded tissues. Indirect evidence from inhibitor and genetic studies indicates that there are probably several different Type III synthetases involved in

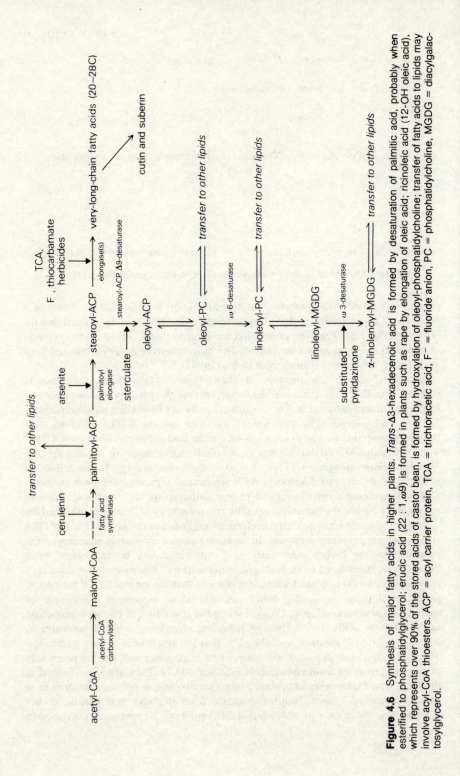

Figure 4.6 Synthesis of major fatty acids in higher plants. *Trans-Δ3-hexadecenoic acid is formed by desaturation of palmitic acid, probably when esterified to phosphatidylglycerol; erucic acid (22 : 1, ω9) is formed in plants such as rape by elongation of oleic acid; ricinoleic acid (12-OH oleic acid), which represents over 90% of the stored acids of castor bean, is formed by hydroxylation of oleoyl-phosphatidylcholine; transfer of fatty acids to lipids may involve acyl-CoA thioesters. ACP = acyl carrier protein, TCA = trichloracetic acid, F⁻ = fluoride anion, PC = phosphatidylcholine, MGDG = diacylgalac-tosylglycerol.*

the elongation of stearate to the series of very-long-chain (up to 30C) fatty acids. These reactions are summarised in Figure 4.6.

The Type II synthetase that is present in the chloroplasts of *Euglena gracilis* grown in the light elongates mainly 14C and 16C acyl chains to stearoyl-ACP, which is probably the precursor for unsaturated fatty acids. The cytoplasmic Type I synthetase produces mainly palmitate, which may be the substrate for very-long-chain fatty acid synthesis in this organism.

In contrast, another unicellular organism, the green alga *Chlamydomonas rheinhardtii*, which possesses a chloroplast and synthesises chlorophyll even when grown in the dark, contains a dissociable Type II synthetase. However, the fatty acids synthesised in the light and dark differ, with longer chain length acids being made in the light. This may reflect the need for more 18C precursors for linolenate synthesis in the light when photosynthesis is occurring.

Although by far the majority of plant unsaturated fatty acids belong to the 18C series, in certain cases significant amounts of very-long-chain unsaturated fatty acids are made. For example, the economically important rape plant (*Brassica napus*) accumulates large quantities of *cis*-13-docosenoic (erucic) acid in its seeds. This is formed by elongation of oleic acid.

Another plant which is rapidly assuming economic importance because its stored lipid is a good substitute for whale oil, is the jojoba bean. The elongation system in this plant utilises oleoyl-CoA as the primer unit with malonyl-CoA as the 2C donor. Surprisingly, if the elongation system is presented with alternative primers, such as palmitoyl-CoA, stearoyl-CoA or stearoyl-ACP *in vitro*, all of these compounds are rapidly elongated in spite of the fact that jojoba only accumulates the *cis*-11-eicosenoic and *cis*-13-docosenoic acids. The explanation for this is that in the intact seed the elongation system has access to the correct oleoyl-CoA substrate only, because of compartmentalisation within the cell.

In contrast to plants (and animals), bacteria do not usually contain significant amounts of fatty acids longer than 18C, so that specific elongation mechanisms (i.e. Type III fatty acid synthetases) are uncommon. A notable exception are the long-chain (up to 56C) fatty acids found in mycobacteria, which are synthesised by a specific ACP-dependent elongation system using medium- or long-chain (depending on the species) thioesters of CoA as the primer instead of acetyl-CoA. The reactions are similar to those of mammalian mitochondria in that acetyl-CoA and NADH are required, and elongation may be related to reversed β-oxidation. Fungi and yeasts have microsomal and mitochondrial elongation systems similar to those of animals, while the green algae resemble plants.

Of all groups in the plant kingdom, marine algae have the greatest variety of *cis*-unsaturated fatty acids. Almost invariably these include large amounts of very-long-chain (20C and 22C) components. Although there have been few studies on the elongation mechanisms involved in their synthesis, the acids have characteristic double bond positions. Thus, major polyunsaturated fatty acids are frequently $18 = 3\omega3$, $18 = 4\omega3$, $20 = 5\omega3$ and $22 = 6\omega3$. Clearly these acids are formed sequentially by a series of desaturation and elongation reactions. It remains to be seen how many different Type III synthetases are involved in such elongations. Judging from the evidence available from animal tissues and with saturated fatty acid elongation in plants, it is extremely unlikely that one enzyme alone is capable of elongating so many different substrates.

4A.3 Fatty acid desaturation

Bacteria have two major pathways of unsaturated fatty acid synthesis. Many members of the Eubacteriales, including all the anaerobes as well as some aerobes and facultative aerobes, possess the 'anaerobic pathway', dealt with above. As its name implies, this pathway does not utilise oxygen and was, perhaps, the evolutionary forerunner of the mechanism used by all other organisms, and many bacterial species, to produce (poly)unsaturated fatty acids, namely aerobic desaturation. The latter process involves the stereospecific removal of two hydrogen atoms from an acyl chain; together with reducing equivalents from NAD(P)H (or other reduced cofactors) that are passed through several redox carriers to the desaturase protein, the hydrogens are used to reduce molecular oxygen to water. With the exception of the soluble stearoyl-ACP desaturase found in photosynthetic tissues, all desaturases are membrane-bound multicomponent complexes that have not been successfully purified from plants or microbes. The basic details of the desaturation process appear to be the same in all organisms in that oxygen and a reduced cofactor are necessary, although there are some differences of detail regarding the nature of the redox carriers (Fig. 4.7), and their sensitivity to inhibition by cyanide. At a mechanistic level there is, in addition, unresolved conflicting evidence from bacterial and algal (or animal) systems as to whether the two hydrogen atoms are removed sequentially (*Corynebacterium diphtheriae*) or simultaneously (*Chlorella vulgaris*).

The position of the first double bond inserted by plants and microbes is, as for animals, usually $\Delta9$. Thus oleate is the product of aerobic desaturation, which clearly distinguishes it in bacteria from the major 18C product of the anaerobic pathway, i.e. *cis*-vaccenic acid ($18 : 1\Delta11$). Bacteria are unique in producing $\Delta10$-monounsaturated fatty acids. Bacilli commonly contain $\Delta5$- or $\Delta10$-desaturases and *Bacillus licheniformis* contains both enzymes if it is grown at low temperatures when it synthesises small amounts of $16 : 2\Delta5,10$ (see Sec. 6A.2). However, although this is polyunsaturated it does not possess the more usual methylene-interrupted structure of those fatty acids. Indeed, a general distinction that is made between bacteria and other organisms is that bacteria are unable to synthesise methylene-interrupted polyunsaturated fatty acids (see Sec. 2A). A particularly interesting exception are the filamentous, gliding bacteria (*Flexibacter* spp.); their major fatty acids are branched, but they also contain considerable amounts of $20 : 5\omega3$, which in some species is the major fatty acid. These organisms are highly motile, but lack flagella, and the large amount of poly-unsaturated (and branched) fatty acids may be necessary to provide a specially fluid cell membrane.

The major fatty acids in plant tissues are unsaturated, particularly those of the 18C series. Plants are unique in that some are capable of forming $\Delta6$-monounsaturated acids (e.g. petroselenic, $18 : 1\Delta6$); animals contain $\Delta6$-desaturases but they only use already unsaturated substrates. Oleic acid in plants is formed from stearoyl-ACP by a $\Delta9$-desaturase, and the enzyme from safflower has been purified by affinity chromatography using columns to which ACP had been linked. The enzyme can use NADPH or photosystems I and II as the source of electrons and the immediate electron carrier is ferredoxin. The use of acyl-ACP as the substrate for desaturation is rare and, apart from one unconfirmed

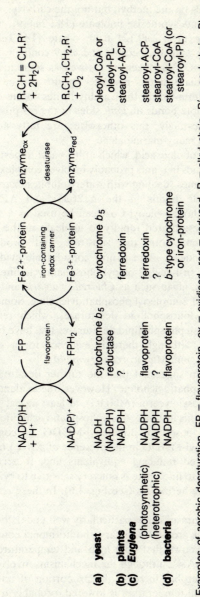

General scheme

$$NAD(P)H + H^+ \qquad FP \qquad Fe^{2+}\text{-protein} \qquad enzyme_{ox}$$

$$R.CH = CH.R' + 2H_2O$$

desaturase

$$NAD(P)^+ \qquad FPH_2 \qquad Fe^{3+}\text{-protein} \qquad enzyme_{red}$$

$$R.CH_2.CH_2.R' + O_2$$

flavoprotein · iron-containing redox carrier

(a) yeast	NADH (NADPH)	cytochrome b_5 reductase	cytochrome b_5	oleoyl-CoA or oleoyl-PL
(b) plants	NADPH	?	ferredoxin	stearoyl-ACP
(c) Euglena (photosynthetic)	NADPH	flavoprotein	ferredoxin	stearoyl-CoA
(heterotrophic)	NADPH	?	?	stearoyl-CoA (or stearoyl-PL)
(d) bacteria	NADPH	flavoprotein	b-type cytochrome or iron-protein	

Figure 4.7 Examples of aerobic desaturation. FP = flavoprotein, ox = oxidised, red = reduced, R = alkyl chain, R' = alkyl chain, PL = phospholipid, ACP = acyl carrier protein.

report in bacteria, seems to be confined to monounsaturated fatty acid synthesis in plants, green algae and *Euglena gracilis*.

In contrast to animals, plants, and most fungi and algae, are able to insert additional double bonds on the methyl, but not the carboxyl, side of a pre-existing double bond. Thus plants synthesise linoleate (18 : 2ω6,9), unlike animals which form octadecadienoate (18 : 2ω9,12) from oleate (18 : 1ω9). Linoleate is an 'essential fatty acid' in animals, because the ω6,9 configuration is necessary for prostaglandin formation. Cyanobacteria (as well as some lower fungi and algae) appear to be intermediate in this respect – groups can be recognised that are animal-, plant- or bacteria-like in their ability to desaturate oleate, and a further group even inserts double bonds on *both* sides of the ω9 position and synthesises 18 : 4ω3,6,9,12. Interestingly, this otherwise rare fatty acid is a major acyl component of lipids in many marine algae.

Unlike the synthesis of oleic acid, which utilises acyl thioesters, the formation of linoleic and linolenic acids in plants probably involves complex lipids. Furthermore, instead of soluble enzymes (dealing with water-soluble substrates) the desaturases introducing *cis*- double bonds in the Δ12(ω6) and Δ15(ω3) positions are membrane bound. Although oleoyl-CoA can be used as a substrate for *in vitro* systems and will be desaturated rapidly to linoleate in the presence of NADH and molecular oxygen, both the substrate and product accumulate in phosphatidylcholine. Indeed, there is now considerable evidence that the actual desaturase substrate is *sn*-l-acyl, 2-oleoyl phosphatidycholine. This lipid has been shown in higher plants, algae such as *Chlorella vulgaris* and several yeasts to be desaturated to *sn*-1-acyl 2-linoleoyl phosphatidylcholine. Some yeasts contain both oleoyl-CoA and oleoylphospholipid desaturases, which respond differently to temperature. To date, all phospholipid desaturases that have been described use an unsaturated substrate, although there is evidence for a stearoyl-phospholipid desaturase in the bacterium *Micrococcus cryophilus*.

In plants it has been suggested that the further desaturation of linoleate to linolenate utilises phosphatidylcholine. However, the evidence at present favours the use of diacylgalactosylglycerol (MGDG), at least in leaf tissues. Linoleic acid must be transferred, in this instance, from phosphatidylcholine to MGDG, and such transfers have been observed. Linoleoyl-MGDG is converted by isolated chloroplasts to linolenoyl-MGDG in the presence of active photosynthesis – which can supply the required reducing equivalents and, if necessary, oxygen. The conversion of linoleate to linolenate is not very sensitive to cyanide and is inhibited by certain pyridazinone herbicides (see Fig. 4.6). In these respects it differs from oleate desaturation.

The control of desaturases is not particularly well understood for any organism. In animals the enzymes are under dietary and hormonal control. In higher plants and green algae it is known that both light and temperature can have profound effects (Secs 6A.2 & 6A.4), although the mechanisms involved are unclear.

Most plants and microbes increase the proportion of unsaturated fatty acids when the environmental temperature is lowered, probably to regulate membrane fluidity (see Secs 6A.2 & 6A.3). This may be brought about by an increase in either enzyme activity or synthesis, or by changes in the levels of reactants. The solubility of oxygen increases as the temperature is lowered, and this fact has been used to explain the increase in unsaturated fatty acid biosynthesis in several plant tissues and in yeast. But the best understood desaturases in terms of their control are those in bacilli, in particular the inducible Δ5-desaturase. Two control mechanisms have been demonstrated for the enzyme in *Bacillus megaterium*: firstly, enzyme synthesis is induced by a decrease in growth temperature, due to the presence of a temperature sensitive modulator which is also responsible for switching off synthesis when the temperature is raised; secondly, the enzyme is thermolabile, so that as the temperature is lowered the half-life of the enzyme increases. In the protozoon *Tetrahymena pyriformis*, the palmitoyl-CoA desaturase appears to be regulated by temperature-induced changes in membrane fluidity, although it is not altogether clear whether enzyme levels alter also. Membrane fluidity also regulates the activity of the Δ9-desaturase in *Micrococcus cryophilus*.

4A.4 Formation of hydroxy fatty acids

Hydroxy fatty acids are formed either as intermediates during various oxidations (see Fig. 4.8 & Sec. 5.B) or by specific hydroxylation reactions. The hydroxyl group is usually at or close to the methyl (ω-end), close to the carboxyl end (α- or β-) or less commonly in the middle of the acyl chain (e.g. 10,16-dihydroxypalmitic acid). Furthermore, 'in-chain' oxidation may yield, particularly in plants, hydroxy, oxo, epoxy, hydroperoxy or polyoxygenated derivatives. A good example of in-chain oxidation in plants is the synthesis of ricinoleic (D-12-hydroxyoleic) acid by castor bean. The enzyme uses oleoyl-phosphatidylcholine as substrate and NADH and oxygen as cofactors and the reaction takes place on the endoplasmic reticulum; it results in the ultimate accumulation of huge quantities of triricinolein – 90% of the dry weight of the seeds!

It is difficult to draw comparisons of the enzymatic mechanisms between animals, plants and microbes because so few systems have been well characterised. Most hydroxylases appear to be mixed function oxygenases in that they require both a reduced pyridine nucleotide and molecular oxygen. Hydroxylases of plants and eukaryotic microbes usually, like those of animals, involve cytochrome P_{450}.

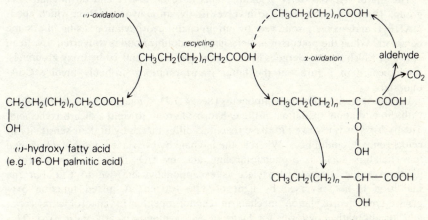

Figure 4.8 Formation of hydroxy fatty acids during oxidations in plants.

Ferredoxin (or a related haem protein) substitutes as the redox carrier in some plant and bacterial systems.

α-Oxidation systems that produce 2-hydroxy fatty acids have been demonstrated in yeasts (e.g. *Hansenula sydowiocum* synthesises 2-OH 26 : 0) and bacteria (e.g. *Arthrobacter simplex* produces 2-OH 16 : 0). The production of 2-OH 16 : 0 has also been shown to occur in plant leaves and cotyledons. In the plant system the D-2-OH 16 : 0 is formed from the D-2-hydroperoxypalmitic acid intermediate of α-oxidation (Fig. 4.8), and such D-OH fatty acids accumulate in plant cerebrosides.

The metabolic origin of the 2-OH and 3-OH lauric and myristic acids present in the lipid A component of the lipopolysaccharide (Secs 2E.3 & 4D.3) of Gram-negative bacteria is unclear. The D-3-OH 12C and 14C fatty acids are normal intermediates of fatty acid synthesis, and it has been proposed that this is the origin of D-3-hydroxy acids in lipid A. However, this cannot account directly for the synthesis of 2-OH acids, nor L-isomers, for which specific hydroxylases or isomerases must exist. For example, in *Pseudomonas ovalis* 3-hydroxyacids arise as by-products of fatty acid synthetase, whereas 2-hydroxy acids are synthesised by a hydroxylation system. This was demonstrated clearly by experiments in which $^{18}O_2$ was incorporated into 2-OH 12 : 0 but not 3-OH 12 : 0.

ω-Hydroxylation systems have been demonstrated in plants, yeasts and bacteria. Those of yeast appear to be very similar to the animal enzymes in that they use NADPH and are cytochrome P_{450}-dependent. This hydroxylation system, like those of several bacterial genera, including *Pseudomonas* and *Acinetobacter*, is also involved in alkane oxidation, a topic of considerable current interest and economic potential. The ω-oxidation system in *Pseudomonas oleovorans* uses NADH and is quite specific for rubredoxin (a non-haem iron-containing protein); NADPH is a poor reductant and cytochrome P_{450} is not involved. The system is composed of three components, viz. rubredoxin, a reductase and the ω-hydroxylase. The 3-hydroxylating system of *Bacillus megaterium*, on the other hand, uses NADPH as electron donor; the main products *in vivo* are 14-OH *iso*- and *anteiso*-branched 15C acids.

The major hydroxy fatty acids in plants have an ω-OH and an in-chain OH group. Their synthesis seems to involve, firstly, an ω-hydroxylation which needs NADPH and oxygen followed by an in-chain hydroxylation using the same cofactors. When the precursor is oleic acid, the double bond is converted first to an epoxide, which then undergoes hydration to give the 9- and 10-hydroxy groupings. As indicated in Figure 4.9 the latter transformations probably involve CoA-thioesters.

For the synthesis of cutin components (Sec. 4D.2), similar reactions are involved with the additional oxidation of the ω-hydroxyl group to yield a dicarboxylic acid. The hydroxylases involved in these reactions differ markedly in their sensitivity to cytochrome P_{450} inhibitors. Whereas the in-chain hydroxylase is inhibited by metal ion chelators such as *o*-phenanthroline and by CO in a reaction which is photoreversible, the ω-hydroxylase is exceptionally sensitive to CO, but the inhibition is not reversed by light of 420–460 nm. A mixed function oxygenase-type hydroxylation mechanism cannot operate in anaerobic microbes and an alternative pathway for hydroxy acid synthesis clearly must exist. This mechanism may be similar to that in the ergot fungus *Claviceps purpurea* or *Pseudomonas* spp., which synthesise 12-OH oleic acid or 10-OH stearic acid by the

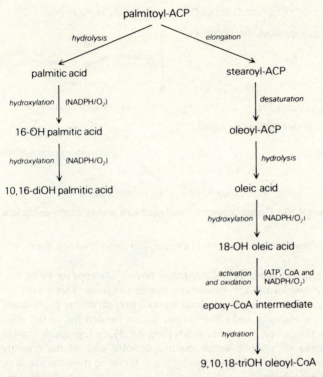

Figure 4.9 Biosynthesis of the major hydroxy fatty acids in plants.

hydration of linoleic and oleic acids respectively – i.e. the hydroxyl group originates from water instead of oxygen. In anaerobic bacteria, of course, the unsaturated fatty acid substrate would originate from anaerobic fatty acid synthetase.

4A.5 Formation of cyclic fatty acids

Cyclopropane fatty acids are formed by the addition of a methylene group from S-adenosylmethionine across the double bond of a monounsaturated fatty acid (Fig. 4.10). This fatty acid is in fact the acyl chain of a phospholipid, so that the substrate *in vivo* is a membrane lipid. The aldehyde residues in plasmalogens and the alcohol residues in alkyl ether lipids in *Clostridium butyricum* also act as acceptors for the methyl group of S-adenosylmethionine to form the corresponding cyclopropane aldehydes and alcohols respectively. The enzyme from *Clostridium butyricum* has been purified and it modifies the fatty acid at the 1-position of phosphatidylethanolamine. Incidentally, this system is interesting because it was the first demonstration of phospholipid acyl-chain modification. The enzyme is not specific for the head-group of the phospholipid, but is relatively specific for the type of unsaturated fatty acids in that it shows a strong preference for ω-7 monoenes (i.e. 16 : 1Δ9 and 18 : 1Δ11), despite the fact that there is more 16 : 1Δ7 than 16 : 1Δ9 in *Clostridium butyricum*. Thus the major cyclopropane fatty acids in bacteria are *cis*-9,10-methylenehexadecanoic acid derived from palmitoleic acid and

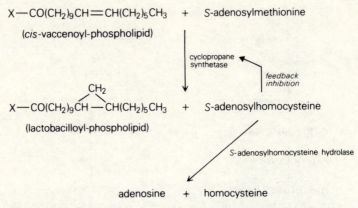

X—CO(CH$_2$)$_9$CH═CH(CH$_2$)$_5$CH$_3$ +   S-adenosylmethionine

(cis-vaccenoyl-phospholipid)

cyclopropane
synthetase

feedback
inhibition

CH$_2$

X—CO(CH$_2$)$_9$CH—CH(CH$_2$)$_5$CH$_3$ +   S-adenosylhomocysteine

(lactobacilloyl-phospholipid)

S-adenosylhomocysteine hydrolase

adenosine   +   homocysteine

Figure 4.10   Cyclopropane fatty acid synthesis and its control in bacteria.

cis-11,12-methyleneoctadecanoic (lactobacillic) acid derived from cis-vaccenic acid.

Although the enzyme from *Clostridium butyricum* appears to be soluble, it is probably released from the membrane during isolation. This seems to occur in *Escherichia coli*, because phospholipid must be present during purification in order to stabilise the enzyme, and phospholipid is also needed for its full activity. Like desaturases that act on membrane lipids (Sec. 4A.3), cyclopropane synthetase must have an active site that is within the hydrophobic core of the membrane. This probably accounts for its loss of activity if it is removed from the membrane in the absence of phospholipid.

In most bacteria the amount of cyclopropane fatty acid depends very much upon the growth conditions and the level generally increases when the growth rate decreases, as in 'step-down' conditions or at the end of the growth cycle in batch culture. The oxygen tension is also important, because the appearance of cyclopropane fatty acids in stationary phase can be prevented by high aeration rates (but see below). At the end of logarithmic growth there is normally little *de novo* phospholipid synthesis and so it is significant that the cyclopropane synthetase uses an intact phospholipid molecule. In *Escherichia coli* the control of cyclopropane fatty acid levels is mediated by changes in enzyme activity since there are no changes in enzyme levels when there is a 30-fold increase in cyclopropane fatty acids. Another point of control is the activity of S-adenosylhomocysteine hydrolase, which modulates the feedback inhibition of the cyclopropane synthetase by S-adenosylhomocysteine (Fig. 4.10). The balance of the activities of the hydrolase and the synthetase may regulate cyclopropane fatty acid levels, which are maximal in *Lactobacillus plantarum* when the two enzyme activities are highest.

In contrast, in *Pseudomonas denitrificans* the amount of cyclopropane synthetase enzyme is under transcriptional control; for example, enzyme synthesis is induced by decreased oxygen tension, and repressed by growth on glucose.

Cyclopropane fatty acids are present in certain species of plants, particularly the Malvaceae and Sterculaceae after which the cyclopropane acids malvalic (cis-8,9-methyleneheptadecenoic) and sterculic (cis-9,10-methyleneoctadecenoic) are named. Cyclopropane acids in plants are made, apparently, by the same mechanism as in bacteria, although the synthetase has not been purified nor have phospholipids been shown directly to act as substrates. The cyclopropene acids are formed by desaturation of the corresponding cyclopropane acid, and malvalic acid is derived from sterculic acid by the removal of the α-carbon atom by oxidation (Fig. 4.11).

Cyclopentenyl fatty acids, such as chaulmoogric acid (13-(2-cyclopentenyl) tridecanoic acid), which has been used for centuries, without much success, in leprosy treatment, are made by the chain lengthening of aleprolic acid using malonyl-CoA:

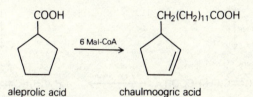

aleprolic acid   chaulmoogric acid

Aleprolic acid itself is probably formed from 2-ketoglutarate via a complicated series of reactions involving the non-protein amino acid cyclopentenylglycine.

ω-Cyclohexyl fatty acids are found in tobacco leaves and in one species of bacterium, *Bacillus acidocaldarius* (where they are probably an adaptation for its hot, acid environment; Sec. 6A.2). The major fatty acids (74–93%) in strains of this organism are 11-cyclohexylundecanoic and 13-cyclohexyltridecanoic acids. These are synthesised by chain elongation of cyclohexyl carboxylate, which is derived from glucose and sugar-phosphates via the first cyclised product, shikimate:

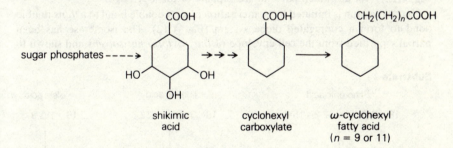

shikimic   cyclohexyl ω-cyclohexyl
acid   carboxylate   fatty acid
(n = 9 or 11)

Thus, both cyclopentenyl and ω-cyclohexyl fatty acids are made by chain elongation of a small cyclic molecule, in contrast with cyclopropane fatty acids which are produced by cyclisation of an already elongated product.

4A.6   Biohydrogenation of fatty acids

Very few organisms are capable of hydrogenating unsaturated fatty acids, and the most common place where they are found is in the rumens of cows, sheep and other ruminant animals. In the rumen there is a large population of anaerobic bacteria and protozoa, some of which (e.g. *Butyrovibrio*, *Eubacterium*, *Fusocillus* and *Borrelia* spp.) can hydrogenate dietary polyunsaturated fatty acids. The predominant acid

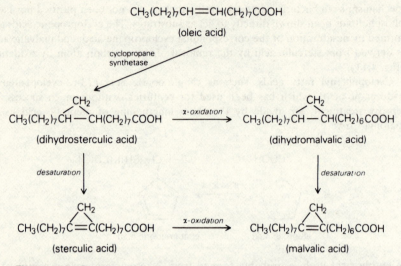

Figure 4.11 Synthesis of cyclopropane and cyclopropene fatty acids in plants.

available is α-linolenic (18 : 3Δ9,12,15 all *cis*) released by lipase action on the chloroplastic glycosylglycerides. The polyunsaturated fatty acids are reduced to a mixture of saturated and monounsaturated *trans* isomers. The hydrogenation of acids such as α-linolenic and linoleic takes place in two stages, namely conversion to a monoenoic fatty acid followed by conversion to stearic acid (summarised in Fig. 4.12). Not all rumen bacteria are capable of this last stage.

Hydrogenation is initiated by isomerisation of a *cis* double bond to a *trans* double bond to form a conjugated diene system (Fig. 4.12). The isomerase has been partially purified from the cell envelope of *Butyrovibrio fibrisolvens* and shown to

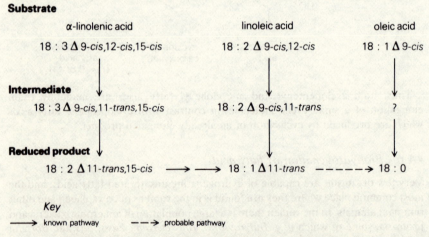

Figure 4.12 Major pathways of hydrogenation in rumen bacteria.

be capable of acting on α-linolenic or linoleic acids, which both contain the necessary *cis*-9, *cis*-12-diene system. The enzyme, in common with all the hydrogenation reactions, also requires a substrate carboxyl group at C1. The mechanistic details of the isomerisation step have not been elucidated, but it is initiated by the stereospecific addition of hydrogen to C13.

The enzyme that hydrogenates the conjugated diene formed by isomerisation has not been isolated. The mechanism is unclear: hydroxy intermediates do not seem to be involved, because in the one case – oleic acid reduction by *Fusocillus babrahamensis* – in which they accumulate, they are not further metabolised. Incorporation studies using D_2O show that hydrogen is incorporated by *cis* addition on the D side of the *cis*-9, *trans*-11-diene, and the reduction may occur by addition of a proton and a hydride ion. The hydrogen donor *in vivo* has not been identified, although experiments with artificial dyes suggest that a low potential redox carrier such as ferredoxin may be involved.

An unusual example of hydrogenation occurs in *Bacillus cereus*, which can reduce oleic to stearic acid. The enzyme is induced by an increase in growth temperature. The regulation of hydrogenation in the rumen flora has not been studied; polyunsaturated fatty acids are more rapidly reduced to monounsaturated fatty acids than are the monounsaturated to the saturated product, stearic acid. Any overall controls in the rumen are likely to be complex, in view of the large and varied microbial population. Although the function of biohydrogenation is unknown, it has been suggested that it may protect organisms from the deleterious effects of polyunsaturated fatty acids on their membranes.

4B   Acyl lipids

4B.1   Triacylglycerols

4B.1.1   Introduction

Triacylglycerols are present as storage lipids in plants, fungi and yeasts, but not bacteria. The biosynthetic pathways in these organisms are similar to those of animals and we shall, therefore, point out only the differences and special points of interest.

4B.1.2   Plants

Plant triacylglycerols are synthesised *de novo* from photosynthetically fixed CO_2. The most widely occurring, and probably quantitatively the most important pathway is the glycerol 3-phosphate (Kennedy) pathway. If fruits of a developing oil-rich plant such as avocado are labelled with [^{14}C]glycerol, the label appears sequentially in glycerol 3-phosphate, phosphatidic acid, diacylglycerol and then triacylglycerol (Fig. 4.13). Although one might expect that glycerol 3-phosphate should be formed from the glycolytic intermediate dihydroxyacetone phosphate, efforts to demonstrate this reaction in sufficient activity in plants have so far proved unsuccessful. However, glycerol kinase is present and would provide an alternative source of glycerol 3-phosphate. The other enzymes necessary (glycerol 3-phosphate acyltransferases, phosphatidic acid phosphatase and 3-acyltransferase) have all been demonstrated in plants. There is some evidence that the 3-acyltransferase may

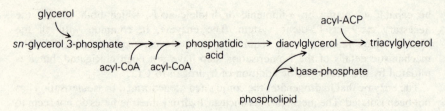

Figure 4.13 Major pathways of triacylglycerol synthesis in plants.

use an acyl-ACP substrate in contrast to the acyl-CoAs of the glycerophosphate acyltransferases.

Besides *de novo* synthesis triacylglycerols can also be formed from phosphoglycerides via diacylglycerol. In plants there is some indirect evidence from labelling experiments for such a pathway. Major phosphoglycerides (in particular phosphatidylcholine) would be converted to diacylglycerol by phospholipase C activity (Sec. 5A.3), or alternatively by a reversal of CDP-choline : diacylglycerol cholinephosphotransferase. In addition to these synthetic pathways, the final fatty acid composition of triacylglycerols is undoubtedly modified considerably by acyltransferase reactions.

4B.1.3 Yeast and fungi

Similar pathways of triacylglycerol synthesis, both *de novo* and from phospholipids via diacylglycerol, are present in yeast and fungi. In fungi, where triacylglycerols may represent 90% of the total lipid, the most important synthetic route is by the acylation of *sn*-glycerol 3-phosphate.

4B.2 Phosphoglycerides

4B.2.1 Plants

Pathways for the formation of phosphoglycerides in higher plants are indicated in Figure 4.14. All the individual enzyme steps share their important details with those from animal tissues and will not be described in detail.

There have been a number of attempts to elucidate control systems for phosphoglyceride synthesis in plants. So far, plant growth regulators such as indoleacetic acid and gibberellic acid have been found to stimulate synthesis (by an increase in enzyme protein). In addition, the nutritional state of the plant is important. For example, in phosphate deficiency a decrease in phospholipid synthesis is (not surprisingly!) seen but other complex lipids such as glycosylglycerides are made at normal rates. Nitrogen deficiency decreases synthesis of phosphatidylcholine, phosphatidylethanolamine and phosphatidylglycerol, but phosphatidylinositol and phosphatidylserine are made at normal rates.

In the last few years, phospholipid exchange proteins have been purified from various animal tissues. Most of these proteins have been found to have a high specificity for one class of phospholipid. In the majority of experiments these proteins catalyse lipid exchange between organelles (e.g. mitochondria and microsomes) but not usually net transfer, which they would have to do *in vivo*. Some phospholipid exchange proteins have also been partially purified from plant

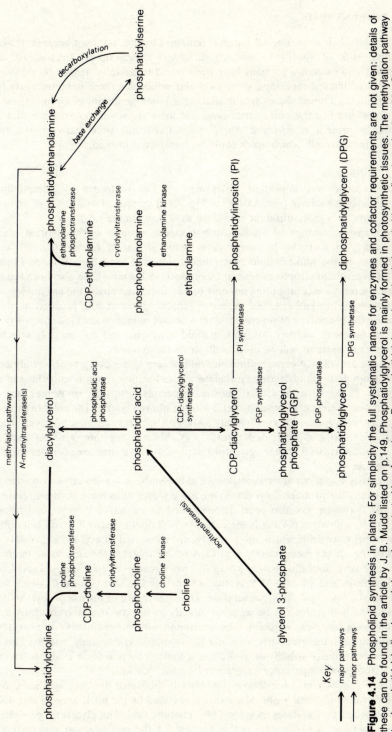

Figure 4.14 Phospholipid synthesis in plants. For simplicity the full systematic names for enzymes and cofactor requirements are not given: details of these can be found in the article by J. B. Mudd listed on p.149. Phosphatidylglycerol is mainly formed in photosynthetic tissues. The methylation pathway to phosphatidylcholine has only been demonstrated in a few plants. Diphosphatidylglycerol synthesis is probably confined to miochondria. Other possible sources of diacylglycerol include phospholipase C action – demonstrated in many plants and plant tissues.

tissues, but their physiological function remains obscure, not least because of their relative lack of specificity towards different phospholipids. In contrast, no phospholipid exchange proteins have been found in bacteria. It may be that with the latters' limited repertoire of intracellular organelles there is no necessity for such proteins. Phospholipids synthesised in the inner membrane of Gram-negatives are transferred to the outer membrane, but there is, as yet, no evidence that an exchange protein is involved; there are 'attachment sites' between the two membranes through which lipids could be transferred instead.

4B.2.2 Bacteria

There are several important differences in the pathways of phospholipid metabolism in bacteria (summarised in Fig. 4.15) compared with those of animals and plants. It is these differences that we shall concentrate on.

A major advantage of studies with bacterial systems is the general ready availability of mutants. In recent years many strains of *Escherichia coli* with mutations in the phospholipid biosynthetic pathways have been isolated (Table 4.4). Glycerol auxotrophs of several Gram-positive bacteria have also been found. These mutants are contributing not only to our understanding of the enzymology of phospholipid biosynthesis, but to wider aspects of membrane function as well, because the phenotype of several involves a change in membrane lipid composition. In the following account of phospholipid synthesis and its control, selected examples of mutants will be used to illustrate these points.

All the enzymes of phospholipid biosynthesis are in the cytoplasmic membrane, apart from phosphatidylserine synthetase whose location remains something of an enigma. In *Escherichia coli* it mostly associates with the ribosomes during purification, whereas in other bacteria it is membrane bound. The soluble form of the enzyme in *Escherichia coli* readily associates with membranes, however, when its K_m for serine is decreased considerably. Several key intermediates such as phosphatidic acid, CDP-diacylglycerol and phosphatidylserine are also found in the membrane.

The early stages of bacterial phospholipid biosynthesis are the same as in animals except that the probable acyl donor for the glycerol phosphate acyltransferase is acyl-ACP (which has also been demonstrated to be active in algae and higher plants); in animals acyl-CoA is the exclusive acyl donor. Many bacteria can utilise exogenous fatty acids which, interestingly, are converted to acyl-CoA (possibly as part of the uptake mechanism). Acyl-CoA is able to act as the acyl donor *in vitro*, but it is very doubtful that it would *in vivo* because of two specific acyl-CoA thioesterases and an acyl-ACP synthetase enzyme in *Escherichia coli*, which also lacks acyl-CoA : acyl-ACP transacylase. The significance of the use of acyl-ACP as acyl donor is that this can be supplied directly by a Type II dissociable fatty acid synthesising complex, which may be associated with the membrane *in vivo*, thus providing an integrated fatty acid and phospholipid synthesising system. This in turn presents the possibility of co-ordinated control; there is some evidence for this from studies of temperature adaptation (see Sec. 6A.2).

There are two acyltransferases involved in phosphatidic acid synthesis, one specific for *sn*-glycerol 3-phosphate that is encoded by the *pls* B gene (Table 4.4), and a second that acylates monoacyl phosphatidic acid. The glycerol 3-phosphate acyltransferase in *Escherichia coli* is specific for the *sn*-1-position and prefers a

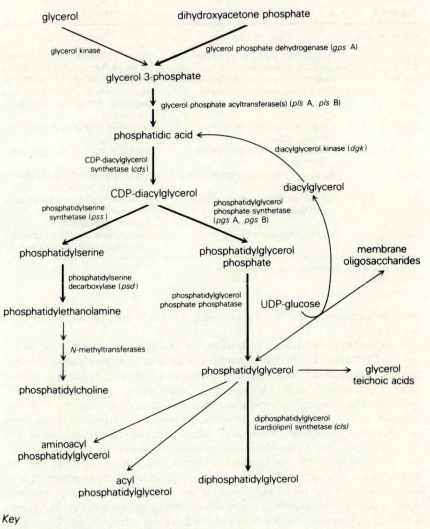

Figure 4.15 Pathways of phospholipid biosynthesis in bacteria. Mutants are indicated in parentheses.

saturated substrate, which partly accounts for the *sn*-1-saturated, *sn*-2-unsaturated acyl composition of the phospholipids. The other factor involved in *sn*-1/*sn*-2 acyl distribution is the ratio of saturated to unsaturated fatty acids produced by the fatty acid synthetase (Sec. 4A.2). The so-called '1-acyl-*sn*-glycerol 3-phosphate route' is also the preferred pathway for phosphatidic acid synthesis in yeasts and animals. In contrast, the phospholipids of mycobacteria are *sn*-1-unsaturated, *sn*-2-saturated, and in this case the preferred pathway is via 2-acyl-*sn*-glycerol 3-phosphate, the

Table 4.4   Mutations of phospholipid biosynthetic genes in *Escherichia coli*.

Name of gene	Enzyme affected	Effect of mutation
gps A	glycerol 3-phosphate dehydrogenase	decreased phospholipid synthesis
pls A	adenylate kinase (ts)	decreased phospholipid synthesis
pls B	glycerol 3-phosphate acyltransferase	decreased phospholipid synthesis
cds	CDP-diacylglycerol synthetase	accumulation of phosphatidic acid
pss	phosphatidylserine synthetase (ts)	decreased phosphatidylserine synthesis
pgs A	phosphatidylglycerol phosphate synthetase	decreased phosphatidylglycerol synthesis
pgs B	phosphatidylglycerol phosphate synthetase	decreased phosphatidylglycerol synthesis
psd	phosphatidylserine decarboxylase (ts)	decreased phosphatidylethanol-amine synthesis
cls	diphosphatidylglycerol (cardiolipin) synthetase	decreased diphosphatidylglycerol synthesis; accumulation of phosphatidylglycerol
dgk	diacylglycerol kinase	diacylglycerol pool size increased, but phospholipids are normal

Abbreviation: ts, temperature-sensitive mutation.

glycerol 3-phosphate acyltransferase being relatively specific for a saturated substrate.

A major difference between bacteria and other organisms is that bacteria do not use the CDP-derivative pathway directly to synthesise phosphatidylethanolamine. Instead, it is produced by the decarboxylation of phosphatidylserine (i.e. formed using CDP-diacylglycerol; Fig. 4.15). In comparison, phosphatidylserine is synthesised in animals almost exclusively by a non-energy-requiring base-exchange reaction. Interestingly, phosphatidylserine decarboxylase is present in the inner mitochondrial membrane, which may reflect the evolutionary origins of this membrane (see Ch. 1). Phosphatidylserine decarboxylase contains bound pyruvate and the enzyme, like the bacterial histidine and *S*-adenosylmethionine decarboxylases, forms a Schiff's base with an amino group in the substrate. It is an extremely active enzyme in most bacteria and is also present in large amounts, more so than is necessary; thus, temperature-sensitive *psd* mutants (Table 4.4) only accumulate phosphatidylserine if the enzyme levels are extremely low. The activity of phosphatidylserine decarboxylase accounts for the virtual absence of phosphatidylserine in Gram-negative bacteria in which phosphatidylethanolamine is usually the major phospholipid. The absence of phosphatidylserine is not due to a low activity of phosphatidylserine synthetase, because this enzyme is also present in excess as demonstrated by *pss* mutants: only in those mutants having almost undetectable levels of decarboxylase is there any increase in phosphatidylserine levels. Furthermore, Gram-positive bacteria such as *Bacillus megaterium*, which contain levels of phosphatidylserine decarboxylase <1% of that *Escherichia coli*, have approximately 6% phosphatidylserine in their phospholipids.

Phosphatidylserine may have a unique role to play in bacteria, because it is present at the N-terminus of several integral membrane proteins. There is evidence that this arises from phosphatidylseryl-tRNA during protein synthesis; alternatively a serine residue at the N-terminus may react with CDP-diacylglycerol in a post-translational modification reaction.

In comparison with other organisms, bacteria lack a CDP-choline pathway and in those relatively few species that contain phosphatidylcholine it is made by sequential methylation of phosphatidylethanolamine using S-adenosylmethionine as the methyl donor. These reactions are similar to those found in yeasts, fungi, plants and animals and will not be discussed further. The above organisms also possess the CDP-choline pathway and in yeast the methylation pathway is suppressed by exogenous choline. It appears that the methylation pathway arose in prokaryotes, and that during the evolution of eukaryotes it was gradually replaced by the CDP-choline pathway leading to a nutritional dependence on choline in higher organisms (which also contain more phosphatidylcholine than bacteria).

The metabolic pathway to phosphatidylglycerol is the same in bacteria as in other organisms, but the mode of synthesis of diphosphatidylglycerol differs from that of animals. In mammalian mitochondria for instance, phosphatidylglycerol reacts with CDP-diacylglycerol to produce diphosphatidylglycerol and CMP. In bacteria diphosphatidylglycerol is synthesised from two molecules of phosphatidylglycerol with the release of glycerol. The lack of involvement of CDP-diacylglycerol in this final step was shown clearly by studies using membrane preparations in which pool levels of CDP-diacylglycerol remained constant; moreover the $^{14}C : {}^{32}P$ ratios of precursor (phosphatidylglycerol) and product (diphosphatidylglycerol) were identical. The enzyme is subject to feedback inhibition by its products. Diphosphatidylglycerol levels often increase at the expense of phosphatidylglycerol under various cultural conditions — e.g. end of exponential growth or sporulation. This may have no physiological function, however, because mutants (cls) that are deficient in diphosphatidylglycerol synthetase and contain 10–50 times less diphosphatidylglycerol (and proportionately more phosphatidylglycerol) show no phenotypic membrane defects.

It is appropriate to introduce a caveat at this point, because most studies of bacterial phospholipid synthesis have been conducted with Gram-negative bacteria, usually *Escherichia coli*. However, there is evidence from experiments with bacilli that there are some differences in phospholipid synthesis in Gram-positive bacteria. The main difference is that phosphatidylglycerol probably acts as a precursor of phosphatidylethanolamine via phosphatidyl exchange. This may, indeed, act as an alternative pathway in *Escherichia coli*, since the reaction can also be demonstrated in serine auxotrophs starved of serine (when phosphatidylethanolamine synthesis from phosphatidylserine is blocked).

Phosphatidylglycerol plays a special role in bacterial phospholipid metabolism, because it is a precursor for several other compounds besides diphosphatidylglycerol (Fig. 4.16). It is perhaps not surprising, therefore, that its turnover rate in *Escherichia coli* and *Bacillus megaterium* is much faster than that of phosphatidylethanolamine, which is normally very stable (cf. Sec. 6A). In *Escherichia coli* it is the sn-1-palmitoyl, sn-2-cis-vaccenoyl molecular species that has the fastest turnover rate compared with disaturated and diunsaturated species, but the physiological significance of this is unclear. The distal glycerol part of the molecule turns over the fastest, possibly due to the reversability of the diphosphatidylglycerol synthetase reaction. A phospholipase D, specific for diphosphatidylglycerol, has also been demonstrated in several bacterial species.

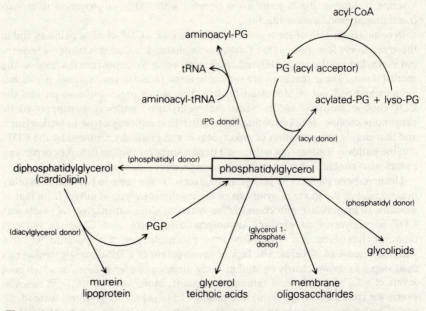

Figure 4.16 Summary of the metabolic interrelationships of phosphatidylglycerol in *Escherichia coli*. PG = phosphatidylglycerol. PGP = phosphatidylglycerol phosphate.

The whole phosphatidylglycerol molecule or different parts of it may act as donor in a variety of reactions, and in the formation of acylphosphatidylglycerol two molecules react, one as donor and the other as acceptor (Fig. 4.16). The role of phosphatidylglycerol in glycolipid biosynthesis is discussed in the next section (Sec. 4B.3). The turnover of phosphatidylglycerol is stimulated when *Escherichia coli* is grown on glucose, due to an increase in the amount of membrane-derived oligosaccharides that are synthesised. These are polymers of glucose substituted with *sn*-glycerol 1-phosphate and phosphoethanolamine residues, the glycerophosphate being derived from phosphatidylglycerol and/or diphosphatidylglycerol. The release of *sn*-1,2-diacylglycerol stimulates the 'diacylglycerol cycle', which is normally a minor route of phosphatidic acid synthesis. Phosphatidylglycerol turnover can also be stimulated by murein lipoprotein (Sec 3B.2.5) synthesis (Fig. 4.16).

The involvement of phosphatidylglycerol and diphosphatidylglycerol in the synthesis of membrane-derived oligosaccharides and murein lipoprotein provides a link between phospholipid biosynthesis and cell envelope assembly and, therefore, cell growth in Gram-negative bacteria. Little is known about the control of the relative amounts of the major phospholipids, although mutants such as *cds* (Table 4.4) may shed light on this because CDP-diacylglycerol represents the branch-point for phosphatidylethanolamine and phosphatidylglycerol synthesis. Phospholipases seem not to be important in this respect. Overall control of phospholipid synthesis may be mediated directly or indirectly through the actions of the guanosine nucleoides ppGpp and pppGpp, which inhibit both fatty acid synthesis (Sec. 4A) and phospholipid synthesis. Their action on phospholipid synthesis is far from fully understood: they inhibit *sn*-glycerol 3-phosphate acyltransferase, but only when acyl-CoA is the substrate, and the *in vivo* significance of regulation by (p)ppGpp remains to be resolved. They are known to regulate RNA and protein synthesis, so that their involvement in phospholipid synthesis would help to co-ordinate all aspects of cell growth. Phospholipid synthesis may also be regulated by the overall energy state of the cell; ATP, and to a lesser extent CTP, inhibit acylation of glycerophosphate. Adenine nucleotides are interconverted by adenylate kinase and so it is significant that the *pls* A mutants, originally isolated as being defective in phospholipid synthesis, in fact have a normal glycerol 3-phosphate acyltransferase but do contain a temperature-sensitive adenylate kinase. The *pls* B mutants, which have a glycerol 3-phosphate acyltransferase with a greatly increased K_m for glycerophosphate, map at a quite different site on the chromosome. In fact the phospholipid synthesis genes are scattered around the *Escherichia coli* genome and no operons have been identified. Although this does not preclude some co-ordinate transcriptional control, it suggests that phospholipid synthesis is more likely to be regulated by metabolic means.

4B.2.3 Fungi and yeast

Phospholipid synthesis in fungi represents an intermediate stage between that in higher plants and that in bacteria. Although phosphatidylethanolamine is made from phosphatidylserine via decarboxylation and can then be converted to phosphatidylcholine by methylation, the CDP-base pathways have also been demonstrated. In general (at least in well studied organisms such as *Saccharomyces cerevisiae* or *Neurospora crassa*) the CDP-base pathways are less important than the 'bacterial' pathways. In yeast, diphosphatidylglycerol is made using CDP-diacylglycerol as an intermediate in the final step, and this nucleotide is also involved in the synthesis of phosphatidylinositol in several, but not all, fungi.

4B.3 Glycolipids

4B.3.1 Introduction

Unlike animals, where glycosylceramides are important, plants contain very small amounts of these compounds and they are not present in bacteria. In contrast, higher plants and algae contain huge amounts of the glycosylglycerides – principally diacylgalactosylglycerol and diacyldigalactosylglycerol (Sec. 2B). Smaller amounts are found in some species of bacteria. These compounds, together with plant sulpholipid, are the characteristic lipids of the photosynthetic membranes of plants and algae. Because of their prevalence they have to be made in enormous amounts on land and in the oceans – approximately 3×10^{14} kg per annum for the three compounds!

4B.3.2 Plants and algae

Not surprisingly, in the eukaryotic cells of higher plants, it is the chloroplasts that are responsible for galactosylglyceride (and sulpholipid) synthesis. Two distinct enzymes are needed for the final steps of galactosylglyceride formation (Fig. 4.17); both are located in the chloroplast envelope and use UDP-galactose which is produced in the cytosol. In contrast, the chloroplast is capable of producing its own fatty acids and diacylglycerol, though it is unclear if it is completely autonomous

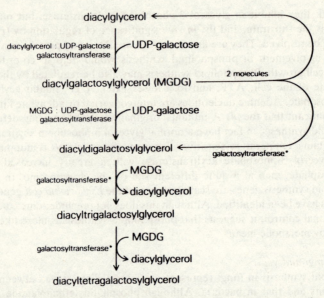

diacylglycerol

diacylglycerol : UDP-galactose
galactosyltransferase — UDP-galactose

diacylgalactosylglycerol (MGDG)   2 molecules

MGDG : UDP-galactose
galactosyltransferase — UDP-galactose

diacyldigalactosylglycerol   galactosyltransferase*

galactosyltransferase* — MGDG
               ↘ diacylglycerol

diacyltrigalactosylglycerol

galactosyltransferase* — MGDG
               ↘ diacylglycerol

diacyltetragalactosylglycerol

Figure 4.17 Synthesis of galactosylglycerides. A single enzyme may perform the three asterisked reactions by transferring a galactose residue from MGDG to the acceptor lipid.

with regard to the huge amounts of linolenate required for galactosylglycerides. The two galactosyltransferases have slight differences in their enzyme characteristics and result in formation of a β-glycosidic and an α-glycosidic bond respectively.

Higher homologues of the galactosylglycerides exist and these are formed by a galactolipid : galactolipid galactosyltransferase (Fig. 4.17). In addition, structural variants that contain glucose instead of one or more of the galactose residues have been found in brown algae and cyanobacteria, and may be more widely distributed.

4B.3.3 Fungi
Glycosylglycerides are common but minor components of fungi – an exception to the latter being *Blastocladiella emersonii*, in which 11–16% of the total lipid is diacyldigalactosylglycerol which is associated with an organelle called the gamma particle that also contains the enzyme chitin synthetase. In addition, fungi produce various extracellular glycolipids (other than sphingolipids) which usually consist of hydroxy fatty acids linked glycosidically or via an ester bond to a carbohydrate moiety. Various glycosyltransferases involved in their synthesis have been described but not studied in detail.

4B.3.4 Bacteria
Gram-positive bacteria contain a variety of glycosylglycerides possessing one to four (usually two) glucosyl, galactosyl or mannosyl residues. Their biosynthetic route is similar to that of plants (Fig. 4.17). The glycosidic bonds are α or β, depending on the organism, and the sugar may vary. As in plants, there are probably distinct transferase enzymes for the formation of mono- and diglycosylglycerides in

bacteria; in *Micrococcus lysodeikticus* the first is membrane bound, while the second is soluble. The enzymes from *Micrococcus lysodeikticus* show a marked preference for substrates containing branched-chain fatty acids with a similar configuration to those found in the glycolipid *in vivo*.

Little is known about the acylated sugar derivatives of bacteria and only the rhamnolipid of *Pseudomonas aeruginosa* has had its biosynthetic route identified. This involves the sequential transfer of two rhamnose units from TDP-rhamnose to β-hydroxydecanoyl-β-hydroxydecanoate.

More is known of the biosynthesis of glycophospholipids, in particular the phosphatidylinositol mannosides. These are formed by transfer of mannosyl residues to phosphatidylinositol; in *Mycobacterium phlei* the donor is GDP-mannose, whereas in *Mycobacterium tuberculosis* it is CDP-mannose. The phosphatidylinositol mannoside may require acylation (at some unidentified site other than the phosphatidyl group) before it will accept a second mannosyl residue. An alternative biosynthetic route involving the transfer of a phosphatidyl group from CDP-diacylglycerol to diacylinositol mannoside has not been confirmed. How the related phosphatidylglycerol glycosides are made is not known. They are unusual in that they contain glucosamine, but the most obvious donor, UDP-glucosamine, has not been found.

There is a close structural and biosynthetic relationship between the diacyldiglycosylglycerols and the phosphoglycolipids of bacteria. For example, it has been shown that the glycerylphosphodiacyldiglucosylglycerol (Sec. 2B.2) of *Streptococcus faecalis* is synthesised by transfer of glycerophosphate from phosphatidylglycerol (or diphosphatidylglycerol) to diacyldiglycosylglycerol. This transfer gives the correct *sn*-glycerol 1-phosphate configuration in the glycerophosphate moiety (compared with *sn*-3 in the phosphatidyl moiety of phospholipids). In comparison, in *Acholeplasma laidlawii* phosphoglycolipids the glycerophosphate moiety has the *sn*-3 configuration, so the biosynthetic route must be different.

4B.4 Plant sulpholipid

The biosynthetic pathway leading to diacylsulphoquinovosylglycerol (the plant sulpholipid) is obscure. Some of the early steps appear to be catalysed by glycolytic enzymes using sulphono intermediates and these may give rise to a nucleoside diphosphatesulphoquinovose which would link with diacylglycerol to form the final sulpholipid. Although the final steps are probably chloroplast-located in plants, it has proved very difficult to obtain active preparations that will catalyse the individual steps. At present, the exact nature of the pathway, including the formation of the critical carbon–sulphur bond in the sulphonate residue, remains an area of much speculation and little evidence!

4C Terpenoids and steroids

4C.1 Pathways to prenols

The early steps of terpenoid and steroid synthesis are common and essentially the same in all organisms. The first important intermediate is mevalonic acid and this is

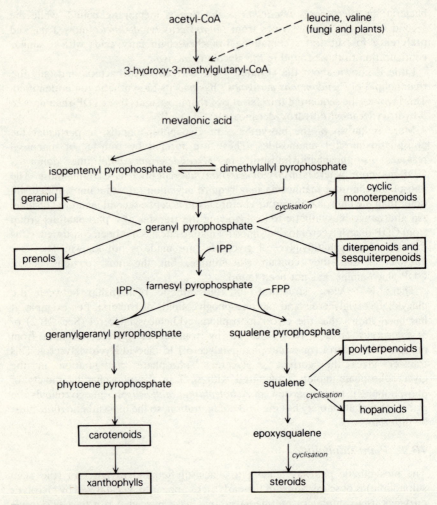

Figure 4.18 Summary of the synthesis of terpenoids and steroids. IPP = isopentenyl pyrophosphate; FPP = farnesyl pyrophosphate.

formed from three molecules of acetyl-CoA (Fig. 4.18). The individual reaction steps are well known and have been studied in detail in yeast and animal systems. The reduction of 3-hydroxy-3-methylglutaryl-CoA (HMG-CoA) is catalysed by HMG-CoA reductase, which is an important regulatory enzyme. In contrast to animals the HMG-CoA of fungi and plants can also be derived from the amino acids leucine and valine, although the relative importance of these sources vis à vis acetyl-CoA is unknown.

Mevalonic acid then gives rise, through a series of three phosphorylations, to isopentenyl pyrophosphate, which possesses the 5C isopentenoid structure characteristic of terpenoids. A crucial feature of isopentenyl pyrophosphate is that, not only does it isomerise to dimethylallyl pyrophosphate, but the two isomers will

isopentenyl pyrophosphate   dimethylallyl pyrophosphate

also condense to give higher oligomers. For example, the two isomers condense to yield geranyl pyrophosphate, the precursor of a range of cyclic and non-cyclic terpenoids, as well as open-chain terpenoid alcohols such as geraniol and the prenols (Fig. 4.8). Many polyprenols of plants (such as the dolichols) have 16–20 isoprenoid units (i.e. 80–100C) and are larger than their bacterial counterparts such as bactoprenol, which is a 55C *cis,trans*-undecaprenol. However, the condensation reactions in plants, animals and bacteria are essentially the same, with the isoprene units being added sequentially, head to tail. At least two types of prenyltransferase enzymes are responsible for the chain elongation reactions – one for *cis* condensations and the other for *trans* condensations. There may also be chain length specific prenyltransferases which determine the length of quinone side-chains (see below) as well as the size of the polyprenol product.

During their involvement as carriers of sugar residues and cell wall precursors, the polyprenol pyrophosphates are released as monophosphates and a specific ATP-dependent kinase rephosphorylates them to the active acceptor pyrophosphate form. In bacteria, this enzyme is inhibited by the antibiotic moenomycin.

Membrane-bound electron transport quinones such as plastoquinones, ubiquinones and menaquinones contain prenyl side-chains attached to a (substituted) *p*-benzoquinone. While geranyl pyrophosphate is a precursor of the prenyl side-chains, the *p*-benzoquinone ring of quinones is derived from shikimic acid or a

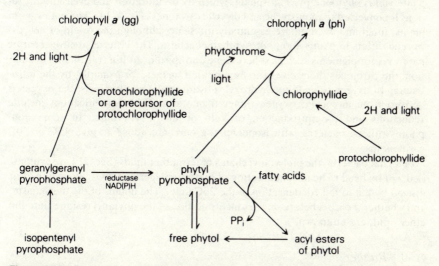

Figure 4.19 Reactions involved in the phytylation step of chlorophyll synthesis. Chlorophyll *a* (gg) represents chlorophyll *a* with geranylgeraniol as the esterifying alcohol; chlorophyll *a* (ph) represents chlorophyll *a* with phytol as the esterifying alcohol; PP_i = pyrophosphate.

related compound. Depending on the number of isopentenoid units that are added, the quinones have a characteristic prenyl chain length and most organisms contain one main type (Sec. 2C). The ring system and prenyl side-chain are linked in a condensation reaction.

Likewise the phytyl chain of (bacterio)chlorophylls is probably attached to the porphyrin nucleus (e.g. chlorophyllide) by esterification using phytyl pyrophosphate (Fig. 4.19), although some doubt remains about the exact route of synthesis. Light is involved in activating the final steps of chlorophyll formation (Fig. 4.19). A possible connection between the esterification of chlorophyllide *a* and the change in structure of the prolamellar body (granal stack precursor) of aetioplasts has been reported. At 0 °C complete photoconversion of protochlorophyllide occurs without any loss of regularity of the prolamellar body. If the temperature is then raised to 26 °C, transformation of the prolamellar body and esterification of chlorophyllide *a* occur concurrently.

4C.2   Carotenoids

Once geranyl pyrophosphate has condensed with a molecule of isopentenyl pyrophosphate to yield the 15C farnesyl pyrophosphate, two metabolic routes are possible. Farnesyl pyrophosphate can either accept another molecule of isopentenyl pyrophosphate to yield the 20C geranylgeranyl pyrophosphate or it undergoes self-condensation to give the 30C squalene. This branch in terpenoid metabolism separates carotenoid from sterol synthesis (Fig. 4.18). In the archaebacteria, which contain 20C phytanyl ether-linked lipids (Sec. 2F), there must also be another branch at this point. All that is known of the latter pathway is that it includes hydrogenation steps to form the saturated phytanyl groups.

The head to head condensation of two molecules of geranylgeranyl pyrophosphate yields phytoene pyrophosphate, which by desaturation and cyclisation reactions is converted to monocyclic and dicyclic carotenes (Fig. 4.20). The pathways in plants, fungi and bacteria are essentially the same, although the origin of neurosporene differs in plants and photosynthetic bacteria. The latter also have separate pathways for pigments such as spheroidene and spirilloxanthin (Fig. 4.20). In addition, the carotenes themselves can be modified further, for example, by the introduction of oxygen to yield the hydroxyl substituents of xanthophylls; the oxygen is derived from molecular oxygen rather than water. Other modifications include reductions and the introduction of keto or methoxy groups. In some non-photosynthetic bacteria extra isopentenyl groups are added to give 45C or 50C carotenoids.

In archaebacteria, the biphytanyl chains of tetraether lipids (Sec. 2F) are synthesised, *not* by head to head condensation of two geranylgeraniol pyrophosphates, but instead by tail to tail condensation of the 20C units. The details of the mechanism, and whether it occurs before or after incorporation of the phytanyl residues into the ether lipid, are unknown.

4C.3   Rubber

Returning to the fate of the open-chain 30C terpenoid squalene, this compound can be elongated further to give the polyterpenes. Only the synthesis of one polyter-

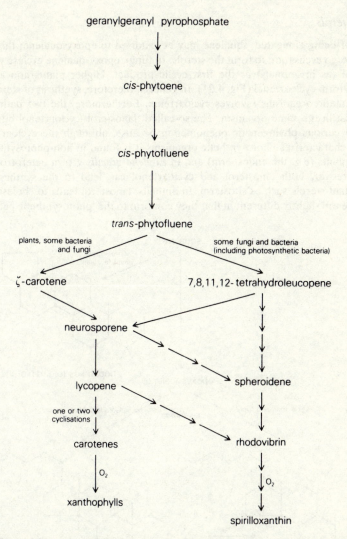

Figure 4.20 Major pathways for the synthesis of pigments in plants and bacteria.

pene – rubber – has been studied in detail. An interesting point to note is that plant latex is living protoplasm. In the latex protoplasm of *Hevea brasiliensis* (rubber tree) rubber particles occupy 30–45% of the volume. The overall production of latex is characterised by the progressive degeneration of organelles comparable to the processes during the development of secretory cells in plant oil glands. In general, the addition of further isoprene units during the synthesis of polyterpenes occurs by the usual condensation process described above. Thus, the starter end of the rubber molecule will bear an isopropylidine group whereas the other end has a pyrophosphate group.

4C.4   Sterols

Instead of being elongated, squalene may be oxidised to epoxysqualene; this can then undergo cyclisation to form the sterols. In fungi, epoxysqualene cyclase yields lanosterol (as in animals) as the first cyclic product. Higher plants and algae, however, form cycloartenol (Fig. 4.21). In general, therefore, synthesis of sterols in photosynthetic organisms involves cycloartenol. Furthermore, the two pathways never exist in the same organism. The so-called 'lanosterol–cycloartenol bifurcation' is a curious phylogenetic phenomenon because, although the cycloartenol pathway characterises photosynthetic organisms it is found in non-photosynthetic parts of plants (e.g. the endosperm) and in *Euglena gracilis* grown heterotrophically. Moreover, both lanosterol and cycloartenol can lead to the synthesis of typical plant sterols such as sitosterol. In animals, lanosterol leads to cholesterol. Red algae are slightly different in that they conform to the 'photosynthetic pattern'

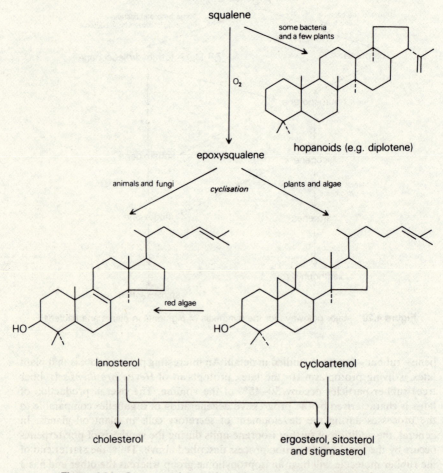

Figure 4.21   Summary of the major pathways to sterols and hopanoids.

in forming cycloartenol. But they then convert this compound via lanosterol to cholesterol, their main sterol. Sterols have been reported in a few cyanobacteria and bacteria but their synthesis has been little studied. Bacteria 'conform' by using the lanosterol pathway.

Cycloartenol contains a cyclopropane ring, in place of the Δ^8 double bond as in lanosterol, and this has to be opened by an isomerase. Before this occurs, alkylation of the side-chain at C24 takes place. Single or double methylations involving S-adenosylmethionine occur. The double bond at C24 is also reduced, a reaction that occurs in plants and animals. However, animals are unable to alkylate at C24, so that the typical plant (and fungal) methylsterols (e.g. campesterol) and ethylsterols (e.g. sitosterol) have a larger side-chain than does for example, cholesterol. Before the final sterol product is obtained, the intermediates still require:

(a) removal of the remaining 4α-methyl group
(b) conversion of the double bond at C7 to one at C5
(c) insertion of a double bond into the side-chain at C22.

Although some of these reactions have been characterised, their sequence and many details are still unknown. In addition, different organisms may use different routes to the same compound – e.g. ergosterol in algae and fungi. Bacteria such as *Methylococcus capsulatus* that contain sterols are unable to carry out the 4α-demethylation and, therefore, contain 4,4-dimethyl and 4-methyl $\Delta^{8(14)}$ sterols. In many fungi, the major sterols are $\Delta^{5,7}$ dienes, for example ergosterol, which suggests that the Δ^7 reductase is absent or almost completely repressed.

4C.5 Hopanoids

In some bacteria and plants, and a few algae (but never in animals), squalene, rather than epoxysqualene, may undergo cyclisation without migration of methyl groups to produce a series of pentacyclic triterpenoids, including the hopanoids (Fig. 2.7) hopene and diplopterol. Hopanoids therefore lack the 3β-hydroxyl group (derived from the epoxy oxygen of epoxysqualene) that characterises the sterols (Fig. 4.21). In contrast to sterol biosynthesis, the formation of hopanoids represents an anaerobic mechanism for the production of related cyclic compounds. Hopanoid derivatives, such as tetrahydroxybacteriohopene, arise by modification of the side-chain, and constituent hydroxyl groups in the side-chain may be substituted with glucosamine. However, the hopanoid nucleus, unlike that of sterols, is never modified.

The ciliate protozoon *Tetrahymena pyriformis* normally incorporates preformed sterols from the environment, but when grown in synthetic laboratory media lacking sterols it synthesises tetrahymenol (Fig. 2.7). As in the synthesis of hopanoids, it is squalene that is cyclised as the starting point for tetrahymenol synthesis. This organism has been used extensively in laboratory investigations of the functions of sterols in membranes and the effects of temperature on membrane lipid composition.

The enzymes responsible for the cyclisation of squalene and of squalene epoxide are quite distinct. Squalene epoxide cyclase from higher organisms is particularly specific, acting only on the (3S)-enantiomer, compared with the enzyme from micro-organisms, which is less specific, leading to the occurrence of 'unusual' compounds such as tetrahymenol.

4C.6 Steroid interconversions

Sterols can be glycosylated by UDP-glucose to form sterol glycosides. The reaction probably takes place in the Golgi complex of active plant tissues. The sugar residues may subsequently be acylated in the presence of a suitable acyl donor (to form acylated sterol glycosides). This occurs in microsomal fractions where such molecules as phosphatidylethanolamine are active donors. Another enzyme is present in soluble preparations from leaves and this uses diacylgalactosylglycerol as the acyl donor. Sterol glycosides are also present in fungi, with sugars such as galactose, mannose and glucose being present.

As an alternative to glycosylation, sterols can be esterified with fatty acids at the 3β-hydroxyl group to yield sterol esters. These compounds are found in plant tissues (Sec. 2D) and in a few fungi. In fungi they accumulate at the end of stationary phase when they may represent as much as 80% of the total sterol content in a few species.

4D Complex lipids

4D.1 Waxes

4D.1.1 Plant waxes

The major components of plant cuticular waxes (Sec. 2E.1) are hydrocarbons, acids, aldehydes, alcohols and esters. All these compounds contain long hydrocarbon chains derived from fatty acids. Depending on the chain length of the constituent, the fatty acid precursor may be a long-chain acid such as palmitate or a very-long-chain fatty acid which would be formed from palmitate by elongation (Sec. 4A.2).

The fatty acid precursor, in the form of its CoA derivative, is reduced by NADH to yield a fatty aldehyde. The latter is then reduced by a second enzyme using NADPH to give a fatty alcohol (Fig. 4.22). An acyl-CoA can then condense with a long-chain alcohol to yield a wax ester. Alternative routes of wax ester synthesis (and less important quantitatively) are by reversal of esterase action or by transacylation involving phospholipid, rather than acyl-CoA, as the acyl donor.

The hydrocarbon components of plant waxes were, for a long time, thought to be made by condensation and decarboxylation reactions. However, they are now known to be synthesised by elongation of the fatty acid precursor until a decarboxylation reaction yields an alkane which is one carbon shorter than the substrate. Both the elongation enzymes and the decarboxylase system appear to be localised in the endoplasmic reticulum. The hydrocarbons, which are generated in the reticulum lumen, are probably excreted via vesicles which fuse with the plasma membrane.

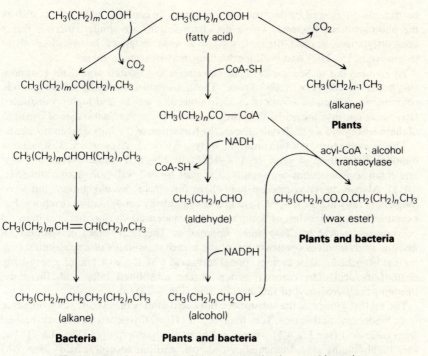

Figure 4.22 Biosynthesis of simple waxes in plants and bacteria.

4D.1.2 Bacterial waxes

There are three types of waxes in bacteria – 'true' wax esters, hydrocarbons and the complex waxes of mycobacteria.

Few bacterial species contain wax esters and little is known about their synthesis. In *Micrococcus cryophilus* the route appears to be via acyl-CoA, as in plants (see above). More is known about the formation of long-chain alcohols, which can be made in two ways. The first route involves oxidation of the corresponding hydrocarbon and generally leads to a secondary alcohol. The second, as in plants, uses the reduction of fatty acyl-CoAs by enzymes which utilise NADH or NADPH depending on the bacterial species (Fig. 4.22). Again, an aldehyde intermediate is formed but it is unclear if two separate enzymes are needed.

A novel function of bacterial fatty acid reductase is in the development of bioluminescence that occurs during the growth of certain species. Light emission by luciferase requires the simultaneous reduction of a long-chain fatty acid to produce the corresponding aldehyde which is reconverted to the acid during luminescence. NADPH is required by the reductase, which is relatively specific for myristic acid. In this case, the fatty acid reductase has been separated from the aldehyde reductase.

Both of the routes of hydrocarbon synthesis mentioned above (Sec. 4D.1.1) have been demonstrated in bacteria. The relative contribution of each route seems to depend on the chain length of the hydrocarbon product. Where the hydrocarbons are much longer than the lipid fatty acids (e.g. in *Micrococcus luteus*), head to head condensation followed by decarboxylation occurs. In contrast, in organisms such as the photosynthetic bacteria, where the chain lengths are similar, there is direct decarboxylation. Several organisms have a wide range of hydrocarbon chain lengths (e.g. 15–31C) and both mechanisms may operate in them.

As pointed out in Section 2E.2, mycobacteria and related organisms contain a vast array of complex 'waxy' lipids. These contain carbohydrate and peptide components as well as a variety of high molecular weight and highly substituted fatty acids, e.g. mycolic acids. Very little is known about the final stages of synthesis of these waxes and we shall only discuss the formation of the fatty acid components.

The mycobacteria contain unusual fatty acid synthetases which synthesise a bimodal distribution of products (14–18C and 22–26C) whose overall amounts and relative proportions are regulated in part by cell wall polysaccharides (Sec. 4A.2). Mycobacteria synthesise long-chain fatty acids by elongation, but very-long-chain fatty acids (32–56C) are probably made by condensation reactions. For example, two molecules of palmitate are condensed during the synthesis of corynomycolic acid in *Clostridium diphtheriae*. The final stages of synthesis of mycolic acids involves condensation of the carboxyl terminus of a substituted (e.g. methyl branched and/or cyclopropane) fatty acid with the α-carbon of a very-long-chain fatty acid. The carboxyl group of the substituted fatty acid, therefore, becomes the β-hydroxyl of the mycolic acid (Fig. 4.23).

The methyl groups in the substituted fatty acid moiety of mycolic acids originate from *S*-adenosylmethionine. Tuberculostearic (10-methylstearic acid) is synthesised from oleic acid (see Fig 4.5) when it is bound to a phospholipid (Sec. 4A.2). The very highly methylated, branched mycocerosic and phthioceranic acids (Sec. 2A) are formed by a different mechanism; the methyl donor is methylmalonyl-CoA for mycocerosic acids, and probably for phthioceranic acids also. In *Mycobacterium tuberculosis* mycocerosic acids are synthesised by an elongation system that uses 18C and 20C primers and methylmalonyl-CoA and is unusual in that it is a soluble enzyme (cf. Sec. 4A.2).

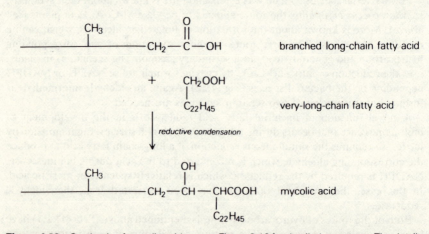

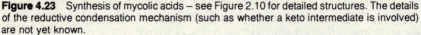

Figure 4.23 Synthesis of mycolic acids – see Figure 2.10 for detailed structures. The details of the reductive condensation mechanism (such as whether a keto intermediate is involved) are not yet known.

Mycobacteria contain a wider range of desaturases than are normally found in other bacteria (e.g. Δ4,Δ5,Δ9- and Δ10-desaturases may be present in the same organism). Combined with elongase enzymes, the desaturases can generate a huge array of positional isomers which can then be modified further by hydroxylation (and oxidation) to keto compounds, methylation, and cyclisation to cyclopropane derivatives. The acids then contribute to the formation of mycolic acids, which consist of families depending on the nature and amount of the substitutions. The enzyme catalysing condensation at the α-carbon (Fig. 4.23) has a restricted substrate specificity and this means that there are fewer types of mycolic acid than predicted from the number of potential substrates.

The antitubercular drug isoniazid appears to act by inhibiting mycolic acid synthesis. In particular, it decreases the amounts of very-long-chain fatty acids (>24C) in mycolic acids and phospholipids. It may block specific elongation systems (cf. inhibition of very-long-chain fatty acid synthesis by thiocarbamate herbicides in plants – Sec. 4A.2) or the Δ5 desaturase, thereby depriving the cell of the correct fatty acids for condensation.

Much less is known about the synthesis of other mycobacterial waxes. The phenolic rings of mycosides A and B are probably derived from propionate via the shikimate pathway, rather than from tyrosine as was thought originally.

4D.2   Cutin and suberin

The complex surface polymers of plants, cutin and suberin, share many components in common (Sec. 2E.1). The synthesis of the constituent molecules, therefore, can be described for both polymers. Palmitic acid, which has been formed by fatty acid synthetase, exists as the ACP-derivative. If unesterified palmitic acid is liberated by thioesterase activity, then it can be hydroxylated at the ω-carbon followed by hydroxylation at C10 (or C7, C8 or C9) (see Fig. 4.9).

If palmitate remains as the ACP-derivative, then it can be successively elongated and desaturated to give oleoyl-ACP (Fig. 4.9). Thioesterase action will then yield unesterified oleic acid which can be hydroxylated in the ω-position in much the same way as palmitic acid was. However, a difference is now seen in the generation of the 16C and 18C families of acids: that is, the ω-hydroxyoleic acid must be

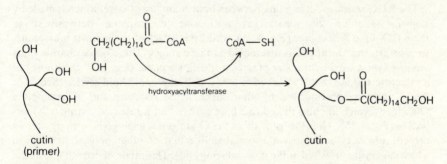

Figure 4.24 Growth of nascent cutin by addition of monomeric units. [Redrawn with permission of the publisher from Kolattukudy, P. E. 1980. In *Biochemistry of plants*, Vol. 4, P. K. Stumpf & E. E. Conn (eds), 571–645. New York: Academic Press.]

activated to its CoA-derivative before an epoxy acid can be formed by a mixed-function oxidase (which uses NADPH and O_2). Hydration of the epoxide then yields hydroxyl groups in the C9 and C10 positions (Fig. 4.9).

The ω-hydroxy fatty acid can also be oxidised to yield a dicarboxylic acid product. An ω-oxo intermediate is involved and the enzyme is very specific for NADP$^+$.

Once the individual monomers have been formed, they are incorporated into the polymer by the action of specific enzymes. Apparently, a CoA derivative of the monomer is required and this is transfered to a free hydroxyl group of the primer molecule. This is illustrated schematically for cutin in Fig 4.24. The transacylase responsible appears to be located in the cuticular matrix – i.e. at the site of cutin deposition.

4D.3 Lipopolysaccharide

Lipopolysaccharide is a complex molecule (Sec. 2E.3) and, predictably, there are several stages in its biosynthesis (Fig. 4.25). The different parts, i.e. lipid A, KDO (3-deoxy-D-mannooctulosonate), R-core and O-antigen polysaccharides are assembled on the inner membrane before transfer of intact lipopolysaccharide to the outer membrane.

The first stage is the synthesis of lipid A, about which least is known largely because mutants defective in its synthesis are invariably lethal. Lipid A anchors lipopolysaccharide in the outer membrane and may be required for the correct functioning of this membrane as well as its assembly. Some information has been gained using temperature-sensitive mutants, such as those of KDO 8-phosphate synthetase (one of the enzymes of KDO synthesis), which synthesise lipopolysaccharide and grow normally below 30 °C but accumulate KDO-deficient precursors in their cytoplasmic membrane at 42 °C when growth ceases. The KDO residues are transferred from CMP-KDO to a partially acylated lipid A, which contains most of its amide- and ester-linked hydroxy fatty acids but no saturated fatty acids, which are added later (Fig. 4.25). The hydroxy fatty acids are added probably as CoA derivatives (cf. the use of acyl-ACPs in phospholipid synthesis; Sec. 4B.2.2). The biosynthesis of lipopolysaccharide hydroxy fatty acids is discussed in Section 4A.4.

The KDO residues act as a link between lipid A and the R-core, which is made by stepwise addition of sugars (as nucleoside diphosphate derivatives) to the KDO–lipid A acceptor (Fig. 4.25), whilst it is attached to the inner membrane. In contrast, the O-antigen is assembled in three stages. The oligosaccharide units are first synthesised by transfer of nucleoside diphosphate (e.g. TDP-rhamnose) to a galactose residue that is attached to a 55C polyisoprenoid lipid carrier (Sec. 2C). These units are polymerised by attachment at the *reducing* end of incomplete O-antigens which are also linked to a lipid carrier that is released during polymerisation (Fig. 4.25). Thus the 'growth' of the O-antigen is analogous to elongation in protein and fatty acid biosynthesis but unlike that of other polysaccharides (in which sugars are attached at the non-reducing end). This mode of growth allows the reducing end of the O-antigen to be held close to the membrane surface where the biosynthetic enzymes are located. Finally, the completed O-antigen is transferred to the R-core, liberating the lipid carrier (Fig. 4.25).

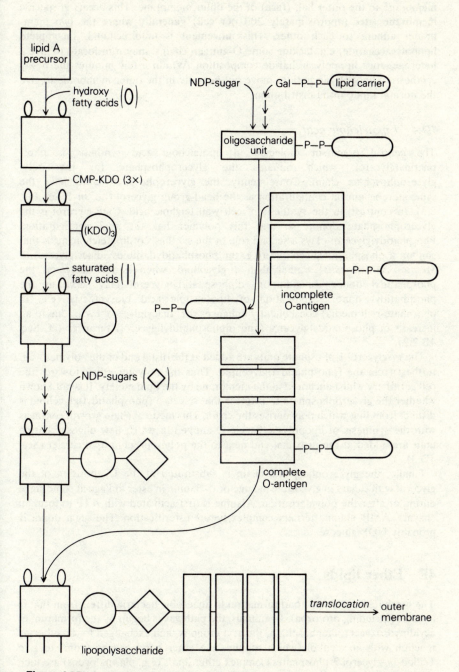

Figure 4.25 A summary of lipopolysaccharide biosynthesis in Gram-negative bacteria.

The last stage in lipopolysaccharide biosynthesis is translocation from the inner membrane to the outer half (face) of the outer membrane. This occurs at specific membrane sites (approximately 200 per cell) generally where the two membranes adhere to each other. This movement is unidirectional. Incomplete lipopolysaccharide, e.g. lacking some O-antigen chains, may translocate to give a heterogeneous lipopolysaccharide composition. Within a few minutes the newly synthesised lipopolysaccharides move out laterally in the outer membrane to give the normal randomised distribution.

4D.4   Lipoteichoic acid

The central precursor molecule in lipoteichoic acid synthesis is phosphatidylglycerol, which donates the glycerophosphate for the polyglycerophosphate chain. Consequently, the glycerophosphate units have the same stereochemical configuration as the head-group glycerol (i.e. *sn*-1; see Fig. 2.3). In contrast, in the synthesis of cell wall teichoic acid, CDP-glycerol is the glycerophosphate donor, so that this polymer has the *sn*-3 configuration. Phosphatidylglycerol plays a second role in the synthesis of lipoteichoic acids that contain a phosphatidylglycolipid (e.g. the phosphatidylkojibiosyldiacylglycerol of *Streptococcus faecalis*), rather than a glycolipid, when it also serves as the phosphatidyl donor. In addition, diphosphatidylglycerol may serve as the phosphatidyl donor. Stimulation of lipoteichoic acid biosynthesis (e.g. in phosphate-rich media) consequently enhances the 'diacylglycerol cycle' due to an increase in phosphatidylglycerol (and diphosphatidylglycerol) turnover (cf. Sec. 4B.2.2).

The *sn*-glycerol 1-phosphate units are added at the distal end of the polymer – i.e. furthest from the (phosphatidyl)glycolipid. Thus the polymer grows towards the cell periphery whilst anchored in the membrane by the lipid moiety. It is not known whether the glycerophosphate transferase that selects a (phosphatidyl)glycolipid is distinct from that which polymerises the chain. This mode of chain growth contrasts with the synthesis of lipopolysaccharide O-antigen in which new oligosaccharide units are added at the proximal end next to the polyisoprenoid lipid carrier (Sec. 4D.3).

Finally, the glycerophosphate chain is substituted at the C2 position of the glycerol with sugars in glycosidic linkage or D-alanine in ester linkage at some stage during or after the polymerisation. Alanine is first activated with ATP to form an enzyme–AMP–alanine ternary complex before esterification. The sugar donor is probably UDP-glucose.

4E   Ether lipids

The mechanism of ether bond formation in anaerobic bacteria differs from that in animals (including protozoa). In animals, the pathway is begun by the formation of acyldihydroxyacetone phosphate; the acyl group is then exchanged for an ether by reaction with an alcohol (which supplies the ether bond oxygen atom) to give 1-alkyl-*sn*-glycerol 3-phosphate. Complex ether lipids (e.g. plasmalogens) are then formed by a series of reactions involving a 1-alkyl-2-acylglycerol 3-phosphate intermediate; the alkyl group is dehydrogenated directly during formation of a plasmalogen.

Such an intermediate does not seem to be involved in the biosynthesis of bacterial plasmalogens. Instead, ether groups are exchanged directly with acyl groups of phosphoglycerides by the use of long-chain aldehydes or alcohols. (The biosynthesis of bacterial long-chain aldehydes and alcohols is covered in Sec. 4D.1.2.)

The second group of bacteria that contain ether-linked lipids are the archaebacteria (Sec. 2F), in which the lipids are based on phytanyl groups linked to glycerol by ether bonds (Fig. 2.14). The origin of the glycerol backbone is through a dihydroxyacetone phosphate intermediate which, unlike that involved in animal plasmalogen synthesis, does not equilibrate with glyceraldehyde 3-phosphate; instead it is presumed to remain enzyme bound, but the mechanism of ether bond formation has not been clarified. The phytanyl groups themselves are synthesised via the mevalonate–isoprenoid pathway (Sec. 4C). Curiously, in extremely halophilic archaebacteria a malonyl-CoA-dependent fatty acid synthesising system does exist, but it is repressed by the high salt concentrations required for growth and no fatty acids are made under normal growth conditions.

Although plasmalogens and other ether lipids are present in plants and fungi, virtually nothing is known about their biosynthesis in these organisms. It is not known, for example, whether the 'bacterial' or 'animal' routes of synthesis are used.

5
Degradation

In a dynamic biological system there exists a fine balance between the activities of synthetic and degradative enzymes. The levels of these activities control the accumulation of individual products, the 'turnover' of most cell components and the utilisation of stored reserves. Thus, catabolic enzymes should not be viewed as mere agents of destruction, but as part of the essential process of life.

The first step in the catabolism of a complex lipid is usually the removal of individual groups, a good example being fatty acids. We shall start our discussion, therefore, with enzymes that attack acyl lipids.

5A   Degradation of acyl lipids

5A.1   Lipases

In the scientific literature there is some confusion as to the term lipase, which is often taken to mean an enzyme that hydrolyses lipids. This is wrong. A true lipase is an enzyme which catabolises triacylglycerols and acts at an oil/water interface. This definition, therefore, precludes enzymes which may cleave a water-soluble, short-chain-length triacylglycerol (esterases) or enzymes which attack other lipids (e.g. acyl hydrolases).

Lipases are widespread in nature and are found in all the phyla. However, high activities are found naturally where there is a major need for the enzymes such as during the germination of oil-rich seeds or the growth of micro-organisms on lipid substrates. The primary ester groups of triacylglycerols are attacked preferentially by lipases which usually attack *SN*-1,2- and *SN*-1,3-diacylglycerols equally although one of the best studied enzymes (that from the fungus *Rhizopus arrhizus*) is specific for the *sn*-1 and *sn*-3 positions. In *Rhizopus arrhizus* the resulting diacylglycerols are hydrolysed more slowly and their product, the *sn*-2-mono-acylglycerol, accumulates during the overall reaction:

$$\text{triacylglycerol} \xrightarrow{\text{fast}} \begin{array}{c} \text{1,2-diacylglycerol} \\ \text{2,3-diacylglycerol} \end{array} \xrightarrow{\text{slow}} \text{2-acylglycerol} \xrightarrow{\text{very slow}} \text{glycerol}$$

In seed tissues two groups of lipases have been reported – 'acid' and 'alkaline' lipases – depending on their pH optima. Although the best studied enzyme, that from castor bean, has a pH optimum of 4.1, other plant seed lipases have alkaline pH optima. The enzymes have different subcellular distributions in different seeds but they all seem to be membrane bound. The high activity of lipases is not only important for the germinating seed, but, incidentally, may cause considerable problems in the commercial use of such seeds.

Many kinds of micro-organism, including bacteria, fungi and yeasts, produce lipases, and some are isolated commercially on a large scale for medical and industrial use – e.g. those with an activity similar to that of pancreatic lipase can be used as a substitute digestive aid in cases of pancreatic insufficiency.

In contrast to plants, bacteria do not store triacylglycerol, and the major function of bacterial lipases is in the breakdown of exogenous triacylglycerol as a food source. The same is true of most yeasts and fungi, although a few species do store triacylglycerol. Examples of where secreted lipases are important include the greasy surfaces of the skin (bacterial and yeast lipases) and in the rumen (anaerobic bacteria); in the soil, bacterial and fungal lipases are responsible for recycling triacylglycerols of plant and animal origin. Most fungal lipases are relatively stable when compared to the enzymes from animal sources, and some are inducible. In certain fungi, the lipases may play a role in spore germination.

Since the majority of microbial lipases are secreted, they are not membrane bound (as in plants) but are soluble enzymes of relatively low molecular weight (20 000–30 000). Some Gram-negative bacteria not only secrete lipases but contain a (probably related) form of the enzyme in the outer membrane. Microbial lipases are usually active over a wide pH range. Some are similar to animal and plant lipases in that monoacylglycerols accumulate. Those of anaerobic bacteria in the rumen differ in that they rapidly convert triacylglycerol to fatty acids and glycerol with no accumulation of intermediates. In ruminant animals fed a cereal and concentrate diet, triacylglycerol is the predominant dietary lipid. But in animals grazing pasture the main lipids are diacylgalactosylglycerol and diacyldigalactosylglycerol. However, the bacterial lipases do not attack these glycolipids (or phospholipids and sulpholipids); instead they are either degraded by other microbial hydrolases to produce mono- and diacylglycerols which are substrates for the lipase or are degraded by acyl hydrolases (see below).

5A.2 Acyl hydrolases

Although enzymes that attack mono- and diacylglycerols, but not triacylglycerols, are known in animals, their existence in plants and microbes is unproven. However, there are enzymes which hydrolyse partial glycerides in marked preference to triacylglycerols – these are known as acyl hydrolases. The acyl hydrolases, as the name implies, actively catabolise a variety of acyl lipids including phospho- and glycosylglycerides. Both fatty acyl groups are hydrolysed and the enzymes can sometimes be extremely active. In the rumen non-specific hydrolases deacylate phospholipids, galactolipids and sulpholipid at very high rates, so that the operation of microbial lipases (as described above) is obviated to a large extent.

In plants the action of acyl hydrolases is prominent in senescent or damaged tissue and, because the enzymes are capable of significant activity at low temperatures, they may cause rotting of incompletely blanched vegetables in a deep-freeze.

In common with most hydrolytic enzymes, acyl hydrolases can act as acyltransferases:

$$R\text{–}\overset{\overset{\displaystyle O}{\|}}{C}O\text{–}X + Y\text{–}OH \rightarrow R\text{–}\overset{\overset{\displaystyle O}{\|}}{C}O\text{–}Y + X\text{–}OH$$

Thus, when YOH is water an unesterified fatty acid is released, but when YOH is methanol a methyl ester is formed. This property, together with the ability of the enzyme to remain active in organic solvents, means that conventional lipid extraction procedures cannot be applied heedlessly to plant tissues without first inactivating the acyl hydrolases by boiling or with hot isopropanol.

5A.3 *Phospholipases*

Radiolabelling studies have shown that the individual parts of phospholipid molecules 'turn over' at different rates. Thus it is obvious that, not only are enzymes present which hydrolyse acyl residues, but there are also others which cleave the base and phosphate. Historically these enzymes have been called by the letters A–D, as shown in Figure 5.1. Phospholipase A is an enzyme which acts on an intact phospholipid molecule at either position 1 (phospholipase A_1) or position 2 (phospholipase A_2). Phospholipase B is in fact a monoacyl phospholipid (lysophospholipid) hydrolase that attacks the product of phospholipase A action.

The phospholipases of snake venoms and in digestive secretions, particularly the phospholipase A_2 from pig pancreas, have been studied widely and various aspects of their catalytic mechanisms are known. Many of these properties, such as the demonstration of the importance of the overall charge on the outside of the lipid micelle substrate, undoubtedly also apply to plant and microbial enzymes. In addition, a distinction should be made between secreted phospholipases, such as the digestive phospholipases of animals and many bacteria and fungi which hydrolyse phospholipid micelles extracellularly, and those that act upon endogenous membrane phospholipids. Apart from the differences in substrate

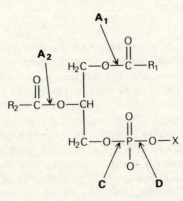

Figure 5.1 The positions of phospholipase action. Note that phospholipases C and D are phosphodiesterases.

configuration, there is the absolute necessity for strict regulation of intracellular phospholipases. This may be achieved by partitioning the enzyme inside lysosomes, but this is not possible, of course, in bacteria. Generally speaking bacterial membrane lipids turn over very slowly, unless the cells are perturbed in some way (e.g. by a sudden shift in growth temperature). A heat-stable protein, which irreversibly inhibited phospholipase A_1, was found in *Bacillus subtilis*; this inhibitor may prevent excessive phospholipid degradation during normal growth.

Phospholipases have been detected in a wide range of bacterial genera, sometimes being associated with their pathogenicity (Sec. 6C.4). The enzymes may be cytoplasmic, membrane bound or secreted. In Gram-negatives they are often found in the outer membrane, and these enzymes may be related forms of those that are secreted. One bacterium may contain several types of enzyme – e.g. *Escherichia coli* contains seven (lyso)phospholipases in various cellular locations and varying widely in their properties.

Phospholipase A is the type most frequently present in bacteria, phospholipase A_1 more often than phospholipase A_2. But in many cases the enzymes are not specific for one position and will also hydrolyse 1- and 2-lysophospholipids – i.e. they are also a phospholipase B. Such an enzyme is the so-called detergent resistant phospholipase A in the outer membrane of *Escherichia coli*, and it is notable for its resistance not only to ionic detergents but temperature also. It is activated by calcium and lacks both acyl chain and head-group specificity. Its physiological role is unclear, especially since *pls* D mutants (which lack the enzyme) have no clear defects in growth or membrane phospholipid metabolism.

A second phospholipase A (i.e. has A_1 and A_2 activity) is found in the cytoplasm of *Escherichia coli*, which in contrast is soluble, thermolabile, sensitive to detergent and relatively specific for phosphatidylglycerol, but does require Ca^{2+}. A number of Gram-negative bacteria have phospholipase A activity in their outer membrane; it is often not determined whether they are phospholipase A_1 and/or A_2. A phospholipase A_1, which like the soluble phospholipase A of *Escherichia coli* is relatively specific for phosphatidylglycerol, is also found in the spores of some bacteria. The spore enzyme is about a thousand times more active than the *Escherichia coli* enzyme and could function in membrane recycling during germination. Curiously, despite the relatively high concentration of Ca^{2+} inside spores, this phospholipase A_1 does not require Ca^{2+} for activity.

The classic phospholipase B comes from the fungus *Penicillium notatum*. At first it was thought that only lysophospholipids could serve as substrates for the enzyme. However, it is now known that normal diacylphospholipids are also catabolised provided that the net charge on the substrate micelle is negative (i.e. the enzyme is really a phospholipase A) – this can be achieved by adding phosphatidylinositol or diphosphatidylglycerol to the other substrate. The *Penicillium notatum* phospholipase also has a low pH optimum of 4.0.

The evidence for intracellular phospholipase C in bacteria is circumstantial. In contrast, several bacteria, particularly pathogens such as *Clostridia* spp., secrete toxins containing phospholipase C (Sec. 6C). These enzymes are Zn-metalloproteins and some also require Ca^{2+} for activation. They vary quite widely in their specificity. Extracellular phospholipase C has been reported in certain phytopathogenic fungi and may be important in their infection of tissues.

There are few examples of phospholipase D in bacteria and their reported activity may be the result of the combined action of several other enzymes. The best characterised is that of *Haemophilus parainfluenzae*, which has a strict requirement for Mg^{2+} and a remarkable specificity for a single phosphate–ester bond in diphosphatidylglycerol between the phosphate and the C-3′ of the central glycerol. The phospholipase D of *Corynebacterium ovis* is specific for sphingomyelin, while the β-toxin of *Staphylococcus aureus* is also a sphingomyelinase.

In contrast, phospholipases A, B and C do not appear to be important in plants, where deacylation of phospholipids is catalysed by non-specific acyl hydrolases (Sec. 5A.2). On the other hand, phospholipase D is widely distributed in higher plants and has broad specificity. It has also been reported in some algae. As with the acyl hydrolase enzymes, phospholipase D is really a transferase enzyme – in this instance catalysing transphosphatidylation:

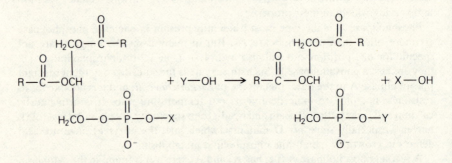

Again this can lead to problems during the extraction of plant tissues with, for example, methanol, when phosphatidylmethanol may appear as a major lipid! Phospholipase D belongs to that group of very active degradative enzymes (including, for example, urease) for which we do not know their natural role. It is found in high amounts in storage tissues and certain leaves such as cabbage. Certainly the enzyme must be essentially inactive *in vivo* – or at least firmly out of harm's way in a specific organelle. A possible function in plants for phospholipase D in synthesising phospholipids has now been largely discounted because the phosphatidylglycerol which is thus made is a racemic mixture unlike the naturally occurring phospholipid (Sec. 2B.2).

5B   Oxidation of fatty acids

The oxidation of fatty acids can be catalysed by a number of enzyme systems which are named after the position of their attack on the acyl chain:

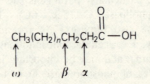

5B.1 α-Oxidation

The process of removing a single carbon atom from the carboxyl end of a fatty acid is called α-oxidation. It is particularly active in plants, but lower activities are also found in bacteria (and animals). The removal of a single carbon atom is important when, for example, the degradation of a fatty acid by β-oxidation (see below) is blocked by the presence of a methyl branch because α-oxidation releases the side-chain as CO_2.

In plants the mechanism of α-oxidation in seeds and leaves is the same. Oxygen and a source of reducing power, e.g. NADH, are required. A non-esterified fatty acid (usually 12–18C) is attacked by molecular oxygen to generate an unstable 2-hydroperoxy intermediate (Fig. 5.2). This intermediate will normally release CO_2 to yield a fatty aldehyde. However, in the presence of enzymes which can reduce peroxides, such as glutathione peroxidase, a D-2-hydroxy fatty acid may be produced which cannot be metabolised easily. Under normal conditions, however, the aldehyde is oxidised to yield a fatty acid (one carbon atom shorter than the original fatty acid) which can then be oxidised further. The cofactor may be a pyridine nucleotide or a flavoprotein depending on the tissue.

In the few bacterial species that have been studied, it appears that α-oxidation also produces a D-2-hydroxy fatty acid. However, in *Escherichia coli* it seems as if this D-form is decarboxylated, as opposed to the situation in plants and animals in which it is the L-form which is decarboxylated preferentially. Thus the hydroxy acids so produced tend not to accumulate in bacteria. The synthesis of lipopolysaccharide hydroxy acids is discussed in Section 4A.4.

In plants, α-oxidation appears to be important for the normal breakdown of fatty acids, as well as the oxidation of branched-chain fatty acids. Except in certain germinating seeds, β-oxidation does not seem to be especially active, and measurements with radioactive substrate almost invariably indicate that α-oxidation is more rapid then β-oxidation. Furthermore, α-oxidation is important in the generation of odd-chain-length fatty acids which are minor components of most plant tissues (cf. bacteria; Sec. 2A). In addition, the 2-hydroxy fatty acid components of certain plant lipids (e.g. cerebrosides) have the D-configuration and may be produced by α-oxidation.

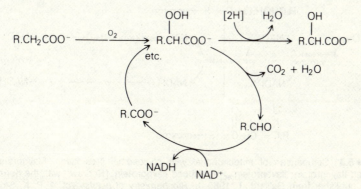

Figure 5.2 α-Oxidation of fatty acids in plants.

5B.2 β-Oxidation

Removal of two carbon atoms from the carboxyl end of a fatty acid is known as β-oxidation. In the microbodies and mitochondria of higher plants, and in the cytoplasm of bacteria, β-oxidation takes place by essentially the same process as has been well described for other organisms such as animals. It will not be discussed further here, therefore, except to point out some differences in plants and bacteria. The overall process of β-oxidation is summarised in Figure 5.3.

As in animals, the fatty acid substrate for β-oxidation in plants must first be activated in the cytoplasm to a CoA derivative and then tranported into the mitochondria. This is achieved by transfer of the acyl group to carnitine. Acyl-carnitines, unlike acyl-CoAs, are able to cross the permeability barrier of the inner mitochondrial membrane. Once inside the mitochondrion, the acyl group is

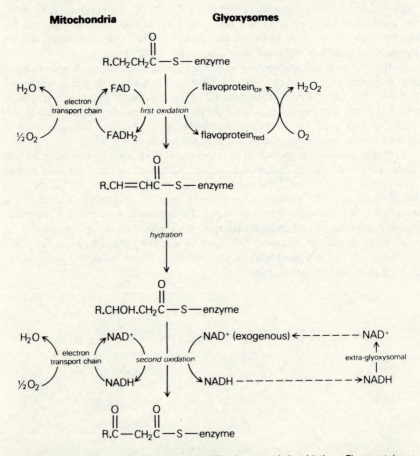

Figure 5.3 Comparison of mitochondrial with glycosomal β-oxidation. Flavoprotein$_{ox}$ = oxidised flavoprotein; flavoprotein$_{red}$ = reduced flavoprotein. [Redrawn with the permission of the publisher from Galliard, T. 1980. In *Biochemistry of plants*, Vol. 4, P. K. Stumpf & E. E. Conn (eds), 85–116. New York: Academic Press.]

returned to CoA. In contrast, bacteria do not contain mitochondria, and β-oxidation takes place in the cytoplasm so there is no necessity for a carnitine transport system.

Bacteria take up unesterified fatty acids from the culture medium by a process that probably involves the formation of acyl-CoA as an integral part of the uptake mechanism, thus making the fatty acid directly available in the correct form for β-oxidation. The so-called fatty acid degradation (*fad*) genes in *Escherichia coli*, coding for the enzymes of uptake/activation and β-oxidation, are located in three sites on the chromosome and comprise a regulon – i.e. they are controlled co-ordinately. Growth on long-chain (i.e. >12C) fatty acids results in the induction of at least five key enzymes. Shorter-chain fatty acids, although substrates for β-oxidation, are not inducers. Several kinds of *fad* mutant are known – e.g. *fad* D mutants lack fatty acyl-CoA synthetase. Because the other β-oxidation enzymes cannot be induced in these mutants, acyl-CoA may well be the natural inducer *in vivo*. Another class of mutants, *fad* R, are constitutive for β-oxidation and are probably repressor mutants. As well as being a repressor for the *fad* genes, the *fad* R gene product may also control the synthesis of the glyoxylate shunt enzymes, isocitrate lyase and malate synthetase. These enzymes, required for the net synthesis of glucose from acetyl-CoA derived from fatty acid oxidation, are elevated in mutants constitutive for β-oxidation.

A somewhat analogous situation with regard to the glyoxylate shunt enzymes exists in a number of oil-rich seeds. Although the mature plants contain the classical scheme for mitochondrial β-oxidation (Fig. 5.3), during the germination of such oil-rich seeds hardly any β-oxidation (less than 3%) takes place in mitochondria. A key organelle here is the glyoxysome, a specialised organelle with a transient existence, whose function is to bring about the efficient mobilisation of stored lipid and its conversion to exportable sugars by the germinating seed (see Sec. 6B.1). This is necessary because, unlike animals, plants do not have a system for transporting fatty acids or triacylglycerols.

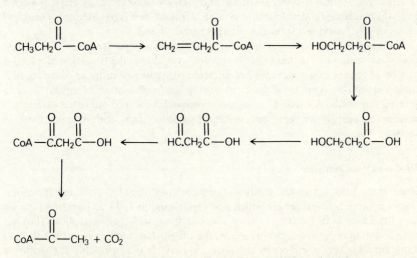

Figure 5.4 Modified β-oxidation in plants for the catabolism of propionate.

The glyoxysome contains all the enzymes of the β-oxidation complex, together with the fatty acid activation enzyme, and these are associated with the glyoxysomal membrane. There are two differences from the mitochondrial β-oxidation system. Firstly, the flavoprotein reduced in the acyl-CoA dehydrogenase step is oxidised directly by molecular oxygen to yield H_2O_2 that is removed subsequently by catalase. Secondly, the NADH produced in the oxidation of α-hydroxyacyl-CoA cannot be re-oxidised in the glyoxysome (Fig. 5.3).

A further difference between plants and animals is that plants have an efficient process for catabolising propionate which differs from the succinic acid pathway used by animal tissues. The plant system involves a hydratase and two dehydrogenase enzymes. In short, it can be termed modified β-oxidation (Fig. 5.4). Although the individual steps are similar to β-oxidation, the fate of the propionate carbon atoms is different.

Microbodies have been found in several yeasts (e.g. *Candida* spp.) which can carry out β-oxidation of fatty acids and contain the enzymes of the glyoxylate pathway. In these species there does not seem to be any β-oxidation activity in the mitochondria.

5B.3 ω-Oxidation

Plants, fungi and bacteria are also capable of oxidising the methyl end of an acyl chain. This is called ω-oxidation and is important either when the carboxyl end is unavailable or for the formation of ω-hydroxy fatty acids (Sec. 4A.4). In addition, oxidation of carbon atoms within the acyl chain can occur, thus forming the various hydroxy, oxo, epoxy, hydroperoxy and polyoxygenated derivatives that are found in nature (see Sec. 2A).

Generally speaking ω-oxidation plays only a minor role in the oxidation of fatty acids or related compounds compared with α- and β-oxidation. This is seen clearly when the metabolism of a detergent, such as alkyl sulphate, is compared in animals and bacteria. Animals usually ω-oxidise and then β-oxidise from the original methyl end, whereas bacteria desulphate to give an alcohol which is oxidised to an acid and subsequently β-oxidised from the original carboxyl end.

Although we have discussed the various oxidations as separate processes, this last example emphasises another important point – viz. that in the oxidation of a lipid it may be necessary to use a number of different oxidations in order to complete acyl chain catabolism. This often happens during the mobilisation of stored lipid in germinating seeds. A specific example is shown in Figure 5.5, where the catabolism of ricinoleic acid (which represents >90% of the stored fatty acids of castor bean) is shown.

5B.4 Lipoxygenase

Some of the most active and easily isolated enzymes often have obscure functions. Such a protein is lipoxygenase which was crystallised in 1947 and about which we have little idea of its function *in vivo*. The enzyme is widespread in plants (but has been found in only one micro-organism, the fungus *Fusarium oxysporum*) and has some very important effects in the food industry. It is responsible for oxidising polyunsaturated fatty acids and pigments such as carotenoids. Indeed, lipoxygenase

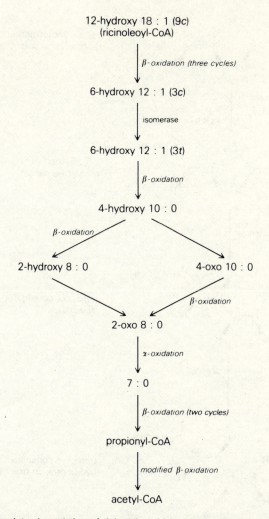

12-hydroxy 18 : 1 (9c)
(ricinoleoyl-CoA)

↓ *β-oxidation (three cycles)*

6-hydroxy 12 : 1 (3c)

↓ *isomerase*

6-hydroxy 12 : 1 (3t)

↓ *β-oxidation*

4-hydroxy 10 : 0

β-oxidation ↙   ↘

2-hydroxy 8 : 0   4-oxo 10 : 0

↘   ↙ *β-oxidation*

2-oxo 8 : 0

↓ *α-oxidation*

7 : 0

↓ *β-oxidation (two cycles)*

propionyl-CoA

↓ *modified β-oxidation*

acetyl-CoA

Figure 5.5 Complete degradation of ricinoleic acid in germinating castor bean. [Redrawn with the permission of the publisher from Galliard, T. 1980. In *Biochemistry of plants*, Vol. 4, P. K. Stumpf & E. E. Conn (eds), 85–116. New York: Academic Press.]

has been used for many years in the baking industry to bleach flour. Furthermore, the destruction of polyunsaturated fatty acids leads to the production of volatile products which form the flavours and aromas (desirable or undesirable!) of many foods. For example, the smell of sliced cucumber is caused by the combined actions of an acyl hydrolase and a lipoxygenase on the endogenous lipids.

Lipoxygenase catalyses the oxygenation of polyunsaturated fatty acids to give hydroperoxides. Although lipoxygenase activity is especially high in the seeds of leguminous plants and certain cereals, it is now clear that the enzyme is generally distributed in higher plants and in various plant organs. In fact, several types of

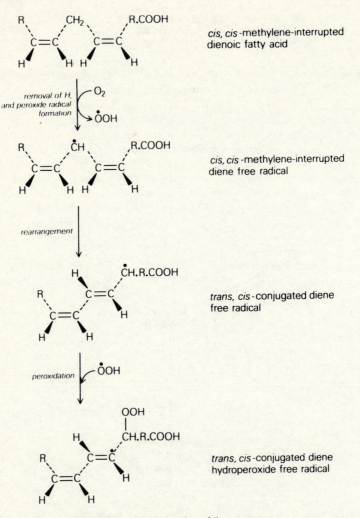

Figure 5.6 Aerobic action of lipoxygenase.ˉ

lipoxygenase with different properties are now known to exist. Most of these enzymes have pH optima in the range 5.5–7.0 while the classic soya bean lipoxygenase I has an optimum of 9.0. The enzymes lack cofactors other than a single atom of non-haem iron; *trans* unsaturated and acetylenic acids are effective inhibitors, as are anti-oxidants or oxygen scavengers. The substrate specificity requires the presence of the *cis*, cis-1,4-pentadiene system found in polyunsaturated fatty acids. Unesterified fatty acids but not their thioesters are utilised by the enzyme. The products of lipoxygenase action are conjugated *cis*-pentadienyl hydroperoxides with the *trans* double bond next to the hydroperoxide. These products are believed to be formed in three steps (Fig. 5.6), and the exact position of the hydroperoxide varies with different enzymes.

Once the aerobic reaction has taken place, lipoxygenase will promote the anaerobic reaction of the hydroperoxide with a molecule of the original polyunsaturated fatty acid. This produces a complex mixture of decomposition products including pentane and oxo acids.

It will be appreciated that hydroperoxides are very toxic and if formed in plants would rapidly damage cellular membranes. Analysis of crude extracts from plant tissues indicates that the hydroperoxides are rapidly reduced to give hydroxy derivatives or converted to ketols or vinyl ethers. In addition, other enzymes can rapidly cleave hydroperoxides to give, for example, volatile aldehydes like hexanal and an oxo acid fragment.

Although the physiological function of lipoxygenase is still unclear, the enzyme may play a role in scavenging oxygen during the germination of certain seeds. However, a wider role may be its involvement with the physiological response to wounding, and in microbial and insect interactions with plants. The active component of traumatin (the wound hormone) is a hydroxyperoxide decomposition product (12-oxo-10-*trans*-dodecenoic acid) and this stimulates the normal wound healing process. In addition, several other products of the action of lipoxygenase (and associated enzymes) have antimicrobial activity or can function as fungal and insect attractants or repellents.

6

Lipid functions

6A   Membrane structure and function

6A.1   Membrane structure

The cytoplasmic membrane of a plant or microbe (just as that of an animal) isolates its cells' contents from the surrounding environment and provides a selectively permeable barrier. The same is true of organellar membranes in plants and eukaryotic microbes, and probably some of the specialised intracellular membranes found in, for example, certain photosynthetic bacteria (although in many the intracellular membranes do not form closed vesicles). In addition, all these membranes provide special, often hydrophobic, environments where enzymes and various cellular processes such as photosynthesis can operate effectively (see Sec. 6A.4). Although it is a commonly held belief that all membranes contain lipid and protein, this does not hold true for all microbial membranes – some halobacteria and cyanobacteria contain gas vacuoles which have a membrane composed of protein and lack lipid entirely.

To carry out their functions cell membranes must be strong and stable enough to maintain their integrity and permeability characteristics, and yet flexibile (fluid) enough to allow conformational changes in membrane proteins, or to withstand gross changes in cell and organelle shape. Our modern concept of membrane structure is identified with the well known fluid mosaic model of Singer and Nicolson. In this, amphiphilic lipids form a fluid bilayer; some proteins (integral) may be embedded in or span the membrane while others (peripheral) may be attached to the surface (facing the cytoplasm usually) (Fig. 6.1). Integral proteins are held by hydrophobic as well as hydrophilic forces to proteins and/or lipids, whereas peripheral proteins interact electrostatically with the lipid head-groups or polar part of other proteins. The alkyl chains of the lipids in the membrane core have a gel to liquid-crystalline transition temperature below the operating temperature of the membrane – i.e. the growth temperature of the organism. The lipids are in a liquid-crystalline phase and the membrane is said to be 'fluid', although the definition and measurement of this property is controversial. What is clear, however, is that this fluidity allows both integral proteins and lipids to move in the plane of the membrane at fast rates (10^{-8}–10^{-11} cm s^{-1}). Thus, for instance, if a phospholipid molecule moved unidirectionally (which it does not, of course) it would traverse the 1.5 μm length of *Escherichia coli* in a minute or so. In contrast, transbilayer movement (called 'flip-flop') is relatively slow, with a half-time of

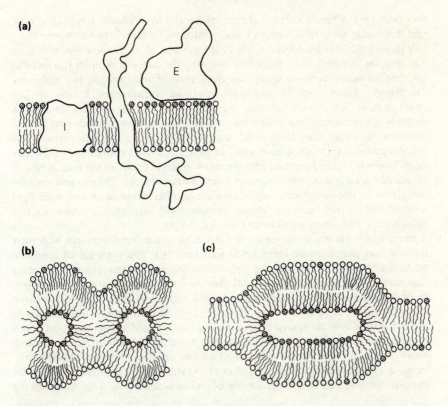

Figure 6.1   Some arrangements of lipids in membranes: (a) classical bilayer with intrinsic (I) and extrinsic (E) proteins; (b) hexagonal-II phase (inverted micelles) sandwiched between a lipid bilayer; (c) fusion of inverted micelles within a bilayer. The asymmetry of lipid arrangements is indicated by the shaded and unshaded head-groups. However, these are not intended to represent particular lipids or their abundances.

minutes; this rate is slow enough to maintain the asymmetry of membrane lipids *in vivo* that is found in plants and bacteria, as in animals (see Sec. 3C).

Although most lipids such as phosphatidylcholine, phosphatidylglycerol, phosphatidylserine and diacyldigalactosylglycerol, adopt a bilayer configuration (lamellar phase), there are certain types, e.g. unsaturated phosphatidyl-ethanolamine, diphosphatidylglycerol and diacylgalactosylglycerol, for which depending on the temperature and the Ca^{2+} concentration, the preferred configuration is hexagonal-II (Fig. 6.1). In hexagonal phases the lipid molecules are arranged in a sphere or a cylinder rather than as a 'flat' membrane. In the hexagonal-I configuration the polar head-groups are on the outside, while in the hexagonal-II configuration they are buried in the interior. It should be noted that hexagonal-II phase-forming lipids are common in bacteria and some other plant and eukaryotic microbial membranes including the chloroplast thylakoids. Non-bilayer regions in membranes composed of bacterial or chloroplast lipids have been visualised by freeze-fracture electron microscopy. Their functions are unclear:

they provide a different balance of hydrophilic and hydrophobic forces (Fig. 6.1), and they could be used to integrate bulky intrinsic proteins into the membrane.

Various estimates have been made of the amount of interaction between lipids and proteins in membranes. Generally speaking the majority of lipids (up to 75%) are free to move without hindrance from protein interactions, but there are exceptions. These include, particularly, those membranes having a low lipid : protein ratio such as the purple membrane of halobacteria. This contains approximately ten molecules only of lipid per bacteriorhodopsin protein molecule; although they form small regions of bilayer in the interstices of the two-dimensional quasi-crystalline array or proteins, the alkyl chains are unusually rigid. This is in marked contrast with the relatively fluid nature of the lipid in animal retinal disc membranes, which contain a very similar protein. The pigment–protein complexes in photosynthetic membranes are another example of restricted lipid motion due in this case to strong protein–lipid interactions which have a considerable functional importance (see Sec. 6A.4).

Much less is known about membrane structure in the archaebacteria, with their unusual and characteristic ether lipids (see Sec. 2F). The presence of saturated ether-linked alkyl chains and the opposite configuration (compared with other organisms) of the glycerol residues may provide stability against attack by oxidation or phospholipases. Archaebacterial cell walls lack peptidoglycan, which in true bacteria imparts rigidity and strength to the cell. At least part of this function in archaebacteria may be served instead by the bidiphytanyl tetraether lipids which span the membrane, thereby locking the two halves of the bilayer relative to each other (Fig. 6.2). The phytanylsesterpenyl diether lipids of thermoalkaliphiles may perform a similar function, and a number of 'zip structures' have been proposed for the hydrophobic core of the membranes of archaebacteria (Fig. 6.2). Although the

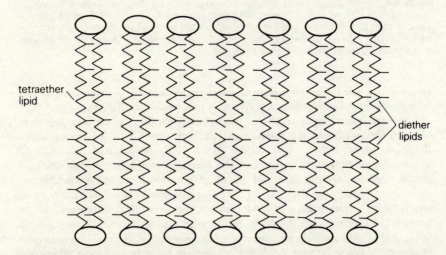

Figure 6.2 Proposed model for the hydrophobic lipid core of archaebacterial membranes containing tetraethers and diethers. [Adapted with the permission of the Society for General Microbiology from De Rosa, M., A. Gambacorta, B. Nicolaus, H. N. M. Ross, W. D. Grant and J. D. Bu'Lock 1982. *J. Gen Microbiol.* **128**, 343–348.]

recently discovered macrocylic glycerol ethers (see Sec. 2F) do not span the membrane, the bridging of the glycerol oxygens by the biphytanyl chains presumably reduces lipid mobility compared with a diphytanyl diether and may impart strength to the membrane. But as yet we know little of the physical properties of archaebacterial lipids and membranes.

The halobacteria, which require 4M salt for growth, contain lipids in a liquid-crystalline state at the usual growth temperature. Salts play a role in the creation of this fluid phase by neutralising the charged head-groups; squalene is also needed to intercalate between the alkyl chains and space the head-groups far enough apart to allow penetration of the cations. When they are grown photosynthetically the red halobacterial membranes develop purple patches, which contain essentially a single protein species – bacteriorhodopsin. The major lipid in both red and purple membranes is phosphatidylglycerol phosphate which *in vitro* appears to require the presence of glycolipid sulphate to form bilayers. However, glycolipid sulphate is not present in the red membrane, which, therefore, may not be in the bilayer phase *in vivo*.

One feature of plant and bacterial membranes which is worth emphasising is that, in comparison with animal membranes, they contain very little sterol. The action of sterols in modulating lipid fluidity and strengthening membranes is of little importance, therefore, in plants and bacteria. Furthermore, with the exception of archaebacteria (see above), mycoplasmas and a few mutants, all microbes and plants contain cell walls which must help to protect the cytoplasmic membrane to some extent.

6A.2   Adaptation to environmental change

Plants and microbes change their membrane lipid composition in response to a wide range of environmental factors. The nature of these changes was discussed in Section 3E and will not be dealt with further here. Instead we shall consider in more detail the effect on membrane structure and function of one of the commonest environmental changes which bacterial and plant membranes have to withstand – i.e. that of temperature.

Temperature affects directly the fluidity of membrane lipids. Plant and bacterial lipids often undergo a liquid-crystalline to gel transition at a temperature within the growth temperature range of the particular organism. Thus a sudden decrease in growth temperature (e.g. an overnight frost) may result in at least some of the lipid becoming ordered. Most natural membranes contain a mixture of lipids so that the phase transition is broad, and as the temperature is lowered those lipids with the highest melting point phase-separate. The melted chains will tend to increase the motion of neighbouring gel-phase lipids and further broaden the transition. Proteins tend to be 'squeezed out' of gel-phase lipid domains, leading to the formation of protein-free 'patches' that can be seen by, for example, freeze-fracture electron microscopy. Somewhat surprisingly this protein patching does not occur in Gram-positive bacterial membranes even at temperatures below the melting temperatures of their major constituent branched-chain fatty acids. That patching is a property of the fatty acids, and not the proteins, is confirmed by the lack of temperature-dependent patching when fatty acid auxotrophs of *Escherichia coli* (the wild-type strain patches normally) are supplemented with branched fatty acids.

Patching probably allows proteins to continue functioning in the face of transient temperature drops or until the fatty acid composition is changed by adaptation.

Changes in membrane fluidity can be monitored by a variety of physical techniques including electron spin resonance and fluorescence polarisation. Using such measurements one can show, for instance, that winter crop plants like winter wheat have membranes which undergo no major change in their phospholipid ordering in the range 0–40 °C. In contrast, phospholipids from the leaves of a summer plant such as kidney bean (*Phaseolus vulgaris*) undergo a phase transition in the range 10–15 °C.

Generally speaking species of plants or microbes that normally live at lower temperatures have membrane lipid compositions that contain lower melting (e.g. more unsaturated) lipids. This is seen particularly when psychrophilic, mesophilic and thermophilic species of the same genus are compared. But the relationship is not a simple one and many exceptions are known. Part of this undoubtedly stems from the complex interactions between proteins and lipids, and the subtle influences that can be induced by, for example, changing the position of fatty acid esterification or the formation of non-bilayer phases.

Although some microbes and plants have partially ordered membranes at the growth temperature, in the majority the chain melting phase transition is complete or nearly so at this temperature. In Gram-negatives there may be differences between the inner and outer membranes; the latter is often less fluid due to its relatively more saturated lipids (see Sec. 3B.2.5). Since there are such precise temperature-dependent changes in lipid composition, it has been suggested that the apparently strict maintenance of lipid fluidity (homeoviscous adaptation) is a prerequisite for correct membrane function. However, it is not a stringent requirement for growth. For bacteria, experiments with organisms such as *Acholeplasma laidlawii*, which requires fatty acid supplements, or with fatty acid auxotrophs of *Escherichia coli* show that a broad spectrum of fatty acid compositions is tolerated. For example, these organisms can grow with up to 80–90% of their lipids in the gel state. There may be differences in the lower or upper growth temperature limits of cultures with grossly modified fatty acid compositions, but under 'normal' circumstances these limits are *not* set by the lipid fluidity. Significantly, the lipid alters the temperature coefficient of growth, and in a mixed microbial flora growth rate becomes an important factor in the success of a species. This fact is more likely to be the explanation of why fatty acid composition is regulated so finely in response to temperature changes. The membrane enzyme activities (see next Section) and in turn the growth rate are optimised by these means.

Similarly the speed of adaptation may be important, and many plants and microbes have mechanisms for adjusting very rapidly to changes in environmental temperature. This seems to apply particularly to microbes, possibly because of their relatively rapid growth or multiplication rate. This adaptation often involves changes in lipid unsaturation (see Sec. 3E), and in many cases the substrate for desaturation is an intact acyl lipid, thus allowing the rapid modification of membrane lipids *in situ*. For example, although bacilli contain branched-chain fatty acids, a number of species respond to a decrease in growth temperature by inducing a desaturase that effects a rapid, short-term adaptation by modifying existing membrane lipids (see Sec. 4A.3). The desaturase is unstable and a longer-term adaptation process takes over; this involves the *de novo* synthesis of more *anteiso*-(relative to *iso*-) branched fatty acids of shorter average chain length, which are incorporated into new membrane lipids. Bacteria that use the anaerobic pathway of unsaturated fatty acid biosynthesis (see Sec. 4A.2) cannot desaturate membrane lipids *in situ* and instead may increase the speed of the adaptation process by increasing the turnover of lipid fatty acids. However, not all bacteria possess such rapid adaptation mechanisms and they may modify their membranes purely by addition synthesis (Sec. 3E.3.2). The reasons for these differences are unclear, but may be related to the constancy or otherwise of the natural environment.

Algae and cyanobacteria are also capable of adjusting their fatty acid unsaturation rapidly in response to shifts in growth temperature. However, in much of their membranes there is in any case an extremely high level of unsaturation, so that the necessity for changes to preserve 'fluidity' is less clear. Indeed, experiments have been performed with *Chlorella vulgaris* and higher plants which show that the amount of unsaturation seems to be a function of the increased availability of oxygen (solubility in water) at lower temperatures. If the oxygen tension is kept constant as the growth temperature is altered, then no change in fatty acid unsaturation is seen. The rather higher degree of unsaturation in the membranes of higher plants also seems to allow most plants to withstand quite large temperature shifts – e.g. daytime temperatures of 40–50 °C and night-time temperatures of 0 °C for typical desert plants.

6A.3   Regulation of membrane-bound enzymes and transport

6A.3.1   Membrane-bound enzymes

A proportion of the lipid in a membrane will be in contact with integral proteins. This lipid has been called 'boundary' or 'annular' lipid and there is a lively controversy about the exact state (including fluidity) of these lipids and the rate of their exchange with the 'bulk' membrane lipid. Lipids associated with proteins may undergo a liquid-crystalline to gel phase transition at lower temperatures than does the bulk lipid. This is an important point because many probes of lipid fluidity give information about the bulk rather than the boundary lipid, and the conclusions drawn from such experiments, therefore, may be erroneous. However, what is not in doubt is that the activity of integral membrane proteins is very often affected by the physical properties of the lipids immediately surrounding them.

The most important feature of lipid structure with regard to membrane fluidity is the acyl composition; the nature of the head-group is generally less important. Since temperature affects directly the motion of acyl chains, it is temperature that is used most often to probe experimentally the relationship between lipids and membrane enzyme function. Much useful information can be obtained by plotting data in the form of an Arrhenius plot (i.e. log k versus $1/T$, where k is the rate constant of the enzyme reaction). The slope of such a plot is $E_a/2.3R$, where E_a is the activation energy and R is the gas constant. Arrhenius plots of soluble enzymes are usually linear, but those of membrane enzymes may instead be smooth curves or have sharp discontinuities. Different enzymes in the same membrane may not give the same type of plot, indicating that different lipids are associated with each enzyme; alternatively, the enzymes may interact differently with the same type of lipid.

The cause of sharp discontinuities has frequently been interpreted as being due to a change in state of the membrane lipids – e.g. a transition from liquid-crystalline to gel phase. This alters the ease with which the enzyme can undergo the conformational changes involved in catalytic activity, which is represented quantitatively by the activation energy. Sometimes the inflexion temperature and activation energy of a membrane enzyme can be correlated not only with bacterial growth temperature but also with the concomitant changes in fatty acid composition and membrane fluidity. One such example is the Δ9-desaturase enzyme of the psychrophilic bacterium *Microccus cryophilus* (Fig. 6.3), which regulates its membrane fluidity by changes in acyl chain length. Similar correlations are seen in plants, but, of course, in chilling-sensitive plants a sudden lowering of growth temperature may lead to a slowing of growth because cells are damaged! However, even in chilling-resistant plants growth rates change quickly, and more recently this has been correlated with

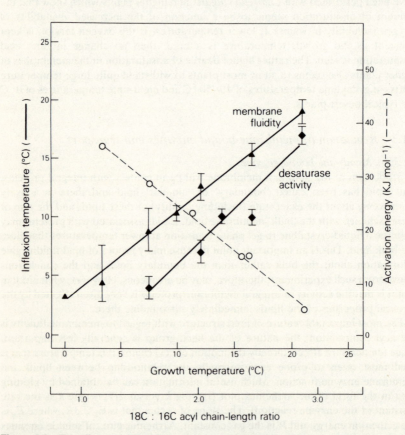

Figure 6.3 The effects of growth temperature and acyl chain length on membrane fluidity and desaturase activity in *Microccus cryophilus*. Inflexion temperatures are obtained from Arrhenius plots of desaturase activity or from plots of $\log_{10}$ (spin label order parameter) versus T^{-1}, where T = absolute temperature; both plots are biphasic with an inflexion at the temperature indicated. The activation energy of the desaturase is derived from the upper slope of the biphasic Arrhenius plots.

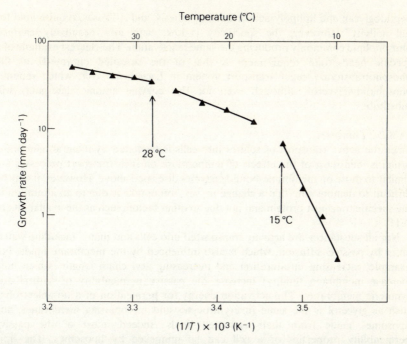

Figure 6.4 Changes in growth of mung bean seedlings as a function of temperature. Seedlings were grown in the dark (when growth energy came from cotyledons). [Redrawn with the permission of the publisher from Raison, J. K. 1980. In *Biochemistry of plants*, Vol. 4, P. K. Stumpf & E. E. Conn (eds), 57–83. New York: Academic Press.]

the transition temperature of membrane lipids (Fig. 6.4). Such correlations based on the physical state of the membrane lipids hold true for a number of enzymes, including several of the respiratory chain enzymes in bacteria and plants, but there are other temperature-dependent parameters that can cause sharp changes in activation energy. These include thermal inactivation of the enzyme, changes in its pH dependency or a change in the K_m (Michaelis constant) of the reaction with temperature. Even when lipid-phase changes do seem to be responsible for discontinuities the actual inflexion temperature may coincide with the start, middle or end of the phase transition (which usually occurs over several degrees Celsius).

The specificity of lipid head-group requirements in lipid–protein interactions has usually been investigated by removing lipids directly with solvents or detergents, or by digesting with phospholipases followed by the re-addition of specific phospholipids. The results are complex and difficult to analyse (particularly after phospholipase treatment when products may stay in the membrane); also, it is sometimes unclear whether the effects could not also be due to differences in fatty acid composition of the various phospholipids. Reconstitution experiments with purified membrane enzymes are equally difficult to analyse because of the problems associated with isolating integral proteins. Notwithstanding such caveats, using these techniques it has been shown that quite a large number of bacterial membrane enzymes, including some of those of respiration, or of phospholipid, peptidoglycan and lipopolysaccharide biosynthesis, and ATPases, require lipid for full activity. However, the specificity is low, with any negatively charged phospholipid commonly producing the same reactivation. The clearest example of a specific head-group requirement is that of the so-called enzyme-II of the phosphotransferase sugar transport system in *Escherichia coli*, which requires phosphatidylglycerol, although even for this enzyme anionic detergents will substitute.

6A.3.2 Transport

Since the active transport of solutes into cells is mediated by integral membrane proteins then most of the effects of temperature on such transport processes are similar to those on membrane-bound enzymes discussed above. However, it may be difficult to demonstrate that a change in, say, ion uptake is due to an alteration in the specific transport protein and not due to other factors such as the availability of ATP.

Not all substances are actively transported into cells and many, including water, enter by passive diffusion, which is also influenced by the membrane lipids. For example, increasing unsaturation and decreasing acyl chain length (which both increase membrane fluidity) increase the passive permeability of natural and synthetic membranes. The activation energy for permeation of a non-electrolyte such as glycerol is the same in erythrocytes and mycoplasma membranes, and liposomes made from their extracted lipids. Indeed, most of the passive permeability properties of a cell can be mimicked by liposomes. The lipid head-group may also affect permeability and this is seen most strikingly in those Gram-positive bacteria that convert a proportion of their phosphatidylglycerol to lysylphosphatidylglycerol in acidic growth conditions. The valinomycin-induced permeability (valinomycin amplifies the natural passive permeability) of both cells and liposomes made with extracted lipids to cations such as Rb^+ is very much less when lysylphosphatidylglycerol replaces the phosphatidylglycerol (Fig. 6.5). This is thought to be an adaptive mechanism to combat the higher proton concentration of the more acidic medium.

Negatively charged lipids such as diphosphatidylglycerol bind divalent cations, particularly Ca^{2+} and Mg^{2+}, which not only increase their availability for transport into the cell but also affect the structure of the membrane: the two electrostatic interactions per molecule condense the lipid bilayer, thus strengthening it. The purple membrane of halobacteria contains glycolipid sulphate, and it has been proposed that this lipid binds K^+ from the environment (which contains low K^+ and high Na^+) for transport to the inside (which contains high K^+), although in this instance the steep concentration gradient is maintained by active transport through a light-driven ATPase rather than by passive diffusion.

The effects of acyl chain composition on passive permeability are also demonstrated by the fact that as liposomes are cooled through the transition temperature of their constituent lipids, the rate of water permeation decreases tenfold and the activation energy increases considerably (e.g. from 36 to 109 kJ mol^{-1}). Both electrolyte and non-electrolyte permeabilities are maximal at the transition temperature due to either the appearance of transient pores as the lipids change phase or dislocations at the boundary of phase domains. Similar properties have been observed in natural membranes. If chloroplast membranes from a

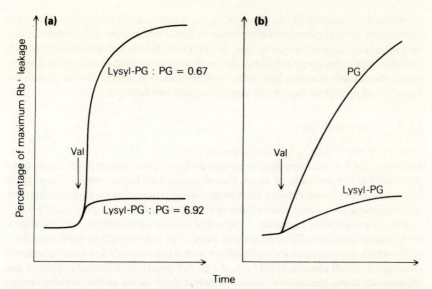

Figure 6.5 The effect of head-group composition on Rb^+ permeability of (a) cells of *Staphylococcus aureus* and (b) liposomes made with extracted lipids. [Adapted with the permission of Dr Haest and Elsevier Biomedical Press BV from Haest, C. W. M., J. de Gier, J. A. F. op den Kamp, P. Bartels and L. L. M. van Deenen 1972. *Biochim. Biophys. Acta* **255**, 720–733.]

chilling-sensitive plant are compared with those from a chilling-resistant plant, it is found that the latter are much less permeable to ions below their transition temperature. In chilling-sensitive plants (e.g. cucumber) not only ions and sugars but even proteins will leak through membranes below their transition temperature. Part of the ability of some plants to withstand moderate freezing temperatures (down to $-10\ °C$) is due to their ability to increase intracellular solute concentrations by allowing water to move out of cells as ice forms on the plant's outside. So long as the transition temperature of such a plant's membrane is well below freezing, then the water permeation is sufficiently fast to allow them to withstand freezing.

Membrane structure and function are also influenced by *raised* environmental temperature. The development of heat injury in plants can be correlated with the thermal lability of enzymes in the chloroplast membranes. Membrane lipid structure appears to be associated with a change in activity of these enzymes at high ($>30\ °C$) temperatures. Photosynthetic activity seems to be much more sensitive to thermal inactivation than are oxidation reactions or membrane transport, and chlorophyll fluorescence is often used as a measure of membrane perturbation. Chlorophyll fluorescence is believed to be a measure of the dissociation of the light-harvesting complex from the reaction centres.

Experiments with bacteria, including some temperature-sensitive mutants, indicate that excessive membrane permeability at elevated temperature does not determine the upper growth temperature limit. For example, although the psychrophilic bacterium *Micrococcus cryophilus* becomes leaky above its optimum growth temperature of 21 °C, the upper growth temperature of 25 °C is due to the presence of several thermolabile aminoacyl-tRNA synthetases. The thermolability of protoplast membranes from the thermophile *Bacillus stearothermophilus* is growth temperature-dependent, but the membranes are always stable a few degrees above the growth temperature. Thus lipid composition affects the thermolability but is not the determinant of the upper temperature limit.

6A.4   Photosynthesis

6A.4.1   Higher plants, algae and cyanobacteria

In Section 3B.1.4, we discussed the unique and invariable nature of the lipids in the photosynthetic membrane of algae, cyanobacteria and higher plants. These lipids obviously confer some definite advantages on such organisms and researchers are just beginning to unravel some of the specific roles which they play (Table 6.1).

The photosynthetic pigments – chlorophylls and carotenoids – are associated with specific proteins. Chlorophyll *a* is present in chlorophyll–protein complex I (CPI) which also contains the reaction centre for photosystem I (and an associated chlorophyll which absorbs at 700 nm, the P–700 pigment). Both chlorophyll *a* and chlorophyll *b* are present (in approximately a 1 : 1 ratio) in the light-harvesting complex (LHC) which is the major thylakoid protein. These complexes (together with chlorophyll–protein complex II) are present in a fluid bilayer which is believed to be composed mainly of the galactosylglycerides (but see below). Other thylakoid acyl lipids probably form a boundary layer of lipids which associate with specific proteins. Thus, diacylsulphoquinovosylglycerol is found within highly purified preparations of CPI while phosphatidylglycerol is present within the LHC. It has been suggested that the unique *trans*-Δ3-hexadecenoate which is found only in the phosphatidylglycerol of chloroplasts is involved in the oligomerisation of LHC which is important in the efficient functioning of photosynthesis. These functions go some way towards explaining the sensitivity of chloroplast membranes to the effects

Table 6.1   Functions of photosynthetic membrane lipids.

Lipids	Properties and functions
chlorophylls (including bacteriochlorophylls)	light-harvesting; electron donor and acceptor at reaction centre
carotenoids	light-harvesting; prevent photo-oxidation of chlorophyll
quinones	electron transport chain components
galactosylglycerides	form bulk of bilayer in most photosynthetic membranes, except those of purple bacteria; diacylgalactosylglycerol readily forms hexagonal phases of unknown function
phosphatidylglycerol	major phosphoglyceride of all photosynthetic membranes; a boundary lipid in chloroplast membranes; associated with pigment–protein complexes (e.g. light-harvesting complexes) in plants, algae and bacteria
diacylsulphoquinovosylglycerol	major component of all photosynthetic membranes, except those of some true bacteria; a boundary lipid; possible functions include associations with chlorophyll and involvement in enzyme activity
ornithine lipids	found only in photosynthetic purple bacteria; may be specifically associated with reaction centre proteins

of lipid hydrolysis. Even a small amount of acyl lipid digestion impairs oxygen evolution and photophosphorylation, although photosystem I-driven electron transport is less sensitive.

Although the major thylakoid lipid, diacylgalactosylglycerol, forms a hexagonal-II phase (Fig. 6.1), it has been suggested that *in vivo* it participates in bilayer formation either by being concentrated in regions of membrane curvature (ends of thylakoids) or because of its association with other membrane components. Formation of hexagonal-II phases *in vivo* may help membrane fusion or protein insertion, and it is possible that these characteristics or other properties are important in photosynthesis. In addition, it has been shown that the polar lipids can associate firmly with chlorophyll and may help to orientate the chlorophyll–protein complexes within the membrane.

6A.4.2 True bacteria
The functional roles of lipids in the photosynthetic membranes of photosynthetic true bacteria, particularly in the purple bacteria, are less clear largely because there are not the same kind of specific lipids as found in chloroplasts; the chlorosomes of green photosynthetic bacteria do contain diaclygalactosylglycerol, but this membrane organelle is probably not the location of the reaction centres (see Sec. 3B.1.4). Chromatophore membranes of some purple bacteria are enriched in phosphatidylglycerol; in others diphosphatidylglycerol or an ornithine-lipid (Fig. 2.6) may be associated with the LHC proteins. Purified reaction centres from photosynthetic bacteria contain very little phospholipid. However, their preparation requires severe treatment with detergents, which replace most of the lipid although a proportion remains tightly bound to the reaction centre proteins – a subunit of the reaction centre in *Rhodopseudomonas sphaeroides*, which is one of the most hydrophobic proteins known, is particularly difficult to isolate free from detergent. The primary photochemical reaction of charge separation performed by the reaction centre *per se* does not require lipid. It is probably the photochemical events involving interaction with antenna bacteriochlorophylls and other proteins that involve tightly-bound lipids.

Recent studies have shown that lipid synthesis in purple bacteria is discontinuous, most lipid being inserted in the chromatophore membrane during the last one-third of the cell growth cycle prior to cell division. Other components are inserted continuously, although it is not clear whether this applies to the reaction centre proteins. Thus, during the cell cycle the lipid : protein ratio of the chromatophore membrane decreases and then increases as division approaches. It remains to be determined how this affects the function of the reaction centres and LHC.

6B Storage

6B.1 Triacylglycerols in plants

The hydrophobic and readily catabolisable nature of triacylglycerols makes them ideal stores of energy in many organisms, including plants. When lipid-rich seeds start to germinate, the triacylglycerol which is stored in oil bodies (see Sec. 3B.1.6)

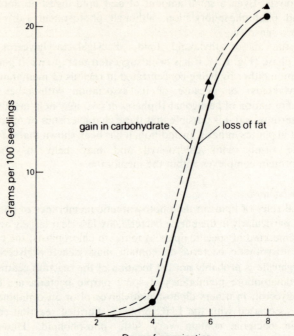

Figure 6.6 Changes in the total amounts of fat and carbohydrate in seedlings of castor bean, *Ricinus communis*, during growth for 8 days at 25 °C. [Redrawn with the permission of the publisher from Beevers, H. 1980. In *Biochemistry of plants*, Vol. 4, P. K. Stumpf & E. E. Conn (eds), 117–130. New York: Academic Press.]

is rapidly broken down. This fat supplies not only energy via respiration but also carbon skeletons for the growing embryo. Thus, in contrast to mammals, fatty seeds need to be able to convert fat into carbohydrate. The importance of this conversion is shown in Figure 6.6. It is brought about by the operation of the glyoxylate cycle (Fig. 6.7), as a by-pass for part of the Krebs cycle. Isocitrate from the Krebs cycle is split by isocitrate lyase to glyoxylate and succinate, and two oxidative decarboxylation steps are avoided. The glyoxylate can then condense with another molecule of acetyl-CoA to give net synthesis of malate, while the succinate can recycle in the remaining part of the Krebs cycle.

The five enzymes special to the glyoxylate cycle are localised in the glyoxysome. This organelle is bounded by a single membrane, and is similar to the mammalian peroxisome in that it contains oxidative enzymes such as catalase, glyoxylate oxidase and urate oxidase. In germinating seeds, the rise and fall in activity of the glyoxylate cycle enzymes is due to changes in the numbers of glyoxysomes. After four days of germination, 20% of the total particulate protein of castor bean seeds is glyoxysomal, whereas at the beginning of germination hardly any glyoxysomes are present. In addition to possessing an active glyoxylate cycle, the glyoxysomes can also activate fatty acids to their acyl-CoA derivatives and they are the only

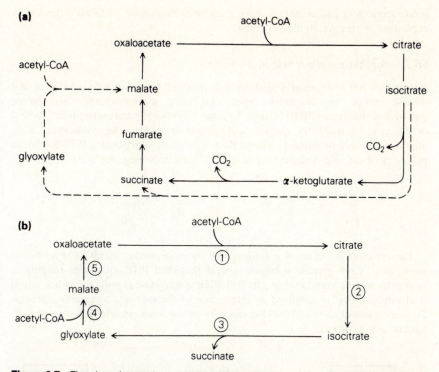

Figure 6.7 The glyoxylate cycle as a bypass of the Krebs cycle (a) and as it operates in the glyoxysome (b). The five steps constituting the cycle are catalysed by (1) citrate synthetase, (2) aconitase, (3) isocitrate lyase, (4) malate synthetase and (5) malate dehydrogenase. In (a) the glyoxylate bypass is shown by the broken lines.

subcellular site of β-oxidation of fatty acids in seeds of plants such as castor bean (see Sec. 5B.2).

Although we have discussed first the importance of the glyoxylate cycle as a provider of carbon skeletons, a major function of stored lipid is to yield energy. The breakdown of triacylglycerol by lipases (see Sec. 5A.1) and the β-oxidation of fatty acids (see Sec. 5B.2) have been discussed previously. The resultant $FADH_2$ and NADH will then be used in oxidative phosphorylation. In unusual plants such as jojoba, where the stored lipid is a wax ester, the hydrolysis of this lipid is followed by oxidation of the released alcohol moiety and β-oxidation of the ensuing fatty acids.

Lipase activity is present in the dry seeds of castor bean and is localised in the oil body membrane. Although lipases are also present in the glyoxysome of such tissues, these enzymes are preferentially active against monoacylglycerols. Thus, seeds are faced *in vivo* with the problem of degrading the fat of oil bodies to fatty acid intermediates which must be transported in some form to the glyoxysomes for further breakdown. Mobilisation of such reserve material is normally controlled hormonally, and removal of the embryo lowers or abolishes the activity of lipases and other enzymes of lipid metabolism. The plant hormones, gibberellic acid, indoleacetic acid and cytokinin, have each been implicated in different tissues as regulators of this metabolic switching.

6B.2 Polyhydroxybutyrate in bacteria

Bacteria do not store neutral lipids such as triacylglycerol, but representatives of a diverse range of bacterial types (including cyanobacteria) accumulate polyhydroxybutyrate (PHB) instead. In some species PHB can make up to 70–80% of the total bacterial dry weight! Such strains are even being considered as an alternative source of starting material for the production of plastics. PHB is a linear polyester of D-(−)-β-hydroxybutyric acid with the following structure:

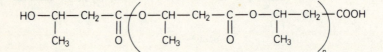

The polymer is stored as granules, 2–8 nm in diameter, which have a limiting membrane. Each granule contains several thousand PHB molecules ranging in molecular weight from 1000 to 250 000. PHB is a crystalline polymer with a helical conformation that is stabilised by interaction of the carbonyl and methyl groups. Polymer configuration and packing can vary giving some granules a 'core and coat' structural appearance.

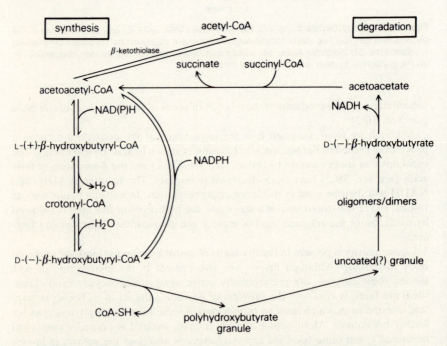

Figure 6.8 The synthesis and degradation of polyhydroxybutyrate in bacteria.

PHB is synthesised when the carbon and energy source is in excess but some other factor, such as N_2 or O_2, is limiting. Different bacteria convert acetoacetyl-CoA to D-(−)-β-hydroxybutyryl-CoA by one of two possible pathways (Fig. 6.8). Although both have acetyl-CoA as the original starting point, they differ in the final step(s) leading to D-(−)-β-hydroxybutyryl-CoA. The final polymerisation step uses CoA-thioesters but not ACP-thioesters, and is catalysed by a synthetase that is bound to the granule.

Degradation of a PHB granule proceeds in three stages (Fig. 6.8). Firstly, the integrity of the limiting membrane is breached by a heat-stable protein whose action is poorly understood. This allows two depolymerases to act; one converts the polymer largely to dimers, and the other hydrolyses oligomers and dimers to the monomer. Finally the D-(−)-β-hydroxybutyrate monomers are degraded to acetyl-CoA, a key step here being the β-ketothiolase reaction that interconverts acetoacetyl-CoA and acetyl-CoA. If Krebs cycle activity is high, the levels of acetyl-CoA are low and CoA-SH is high, and the latter inhibits β-ketothiolase. When nutrient depletion occurs acetyl-CoA rises and CoA-SH falls relieving the inhibition of β-ketothiolase and switching on PHB synthesis. Another control may operate under O_2 limitation when NADH levels rise and Krebs cycle activity is inhibited; this channels acetyl-CoA into PHB synthesis. Storage of PHB uses up excess reducing equivalents (as NADH or NADPH), which can be released as NADH later when needed by catabolism of the granules (Fig. 6.8).

6C Microbial lipids as virulence factors

6C.1 Lipopolysaccharide

It has long been recognised that lipopolysaccharide in the outer membrane of the Gram-negative cell envelope plays an important role in the virulence and survival of pathogenic species. This was first seen in studies of so-called 'rough' (non-virulent) and 'smooth' (virulent) strains of pneumococci. Rough strains lack part or all of the O-antigen polysaccharide side-chains, their increased hydrophobicity rendering them more susceptible to phagocytosis. Although changes in the polysaccharide portion of lipopolysaccharide are often seen and determine the fate of invading bacteria in the body, the whole structure is important in this regard. The lipid A part of lipopolysaccharide activates the alternative pathway of complement fixation (this interaction may be modified by the O-antigen), which leads either to opsonisation and subsequent phagocytosis, or directly to complement-mediated lysis. Lipopolysaccharide acts as a general immunostimulant and, for instance, the inflammatory action of dental plaque has been attributed to bacterial immunogens from intact or degraded lipopolysaccharide (or lipoteichoic acid; see below).

Moribund or dead Gram-negative bacteria release lipopolysaccharide, usually as a complex with protein; this complex is known as endotoxin. The polysaccharide part of lipopolysaccharide and the protein are antigenic, while lipid A is responsible for toxicity. Even very small amounts (a few micrograms) of lipopolysaccharide injected into an experimental animal can lead to fever, local haemorrhage or even death, most of these effects stemming from damage to small blood vessels caused by a hypersensitivity reaction to the lipopolysaccharide.

6C.2 Lipoteichoic acid

The lipoteichoic acid of Gram-positive bacteria possesses most of the biological activities of the lipopolysaccaride of Gram-negatives, including the ability to stimulate the immune system and activate the alternative pathway of complement fixation.

Apart from these effects on the host defence system lipoteichoic acid has beneficial functions for the bacterial cell itself. The polyanionic nature of lipoteichoic acid means that it binds divalent cations at the membrane surface (as do teichoic acids in the cell wall). Esterification with D-alanine (Fig. 2.13) neutralises the anionic nature of the phosphodiester links and regulates the chelating capacity of the polymer. Micrococci which lack lipoteichoic acid have, instead, a succinylated lipomannan that also chelates cations. Lipoteichoic acid has an additional function in many Gram-positives: it regulates endogenous autolytic activity by inhibiting the enzymes that degrade the cell wall peptidoglycan. Only acylated lipoteichoic acid has this function: autolysis-defective mutants contain more acylated polymer (and diphosphatidylglycerol) in their membrane.

6C.3 Mycobacterial cell wall glycolipids

During a search for the virulence agent of pathogenic bacilli in 1950 Hubert Bloch discovered cord factor. Under certain growth conditions virulent strains of mycobacteria grow as long strands ('cords'). Extraction of these with petrol gives a preparation known as cord factor – this is toxic to animals. Chemically, cord factor is a mixture of trehalose dimycolates (see Sec. 2E.2). Its glycolipid nature accounts for many of its biological properties, because it interacts strongly with host cell membranes, such as those of mitochondria and lysosomes, damaging them and, for example, inhibiting respiration.

Cord factor itself is not antigenic but probably acts as a hapten; it binds to albumin in the plasma to form an antigen. Like many other extracts of mycobacterial cell walls, cord factor acts as an immunostimulant; the reactions in different animals can vary considerably. In the laboratory advantage is taken of this immunostimulant property for increasing the titre of antibodies in immunised experimental animals. Killed mycobacteria, or their cell walls, are suspended in paraffin oil (to give Freund's adjuvant) and emulsified with an aqueous solution of the antigen before injection. This increases considerably the yield of antibody. The attenuated bovine strain of *Mycobacterium tuberculosis* (Bacille Calmette-Guerin, BCG) is used for immunisation against tuberculosis in humans.

Levels of sulpholipid in mycobacterial cell walls (Fig. 2.11) have been correlated with the virulence of strains. Sulpholipid inhibits degranulation in polymorphonuclear neutrophils, thus permitting engulfed mycobacteria to avoid intracellular destruction.

Although not related to virulence, microbial glycolipids may have other important properties for the producing organism. For example, the sophorolipids (acylated disaccharides) secreted by the yeast *Torulopsis bombicola* may act as surfactants, solubilising hydrophobic foodstuffs.

6C.4   *Phospholipases*

As discussed in Section 5A.3, bacteria produce a wide range of different phospholipases. Many of these extracellular enzymes are components of toxins (cf. many snake venoms also contain phospholipases). For example, in pathogens such as *Clostridium perfringens* (which is one of the causative organisms of gas gangrene) they form part of the enzyme battery with which the host tissues are broken down. Phospholipase C is particularly active in such toxins. Although produced by many pathogens, phospholipase synthesis is not necessarily directly related to virulence – for example, avirulent strains of gonococci still secrete phospholipase. When present on the surface of non-pathogens, phospholipases can, presumably, act as a nutrient producer.

6C.5   *Siderophores*

Most micro-organisms have to scavenge for iron, since the element has such low solubility at neutral pH and the majority of dissolved iron is in bound form. Bacteria have overcome this problem by producing iron-binding compounds of their own, known as siderophores. These have amazingly high binding constants – e.g. 10^{52} for enterochelin of enteric bacteria! Most siderophores are secreted into the medium where they chelate iron; they are then re-absorbed into the bacterium where they give up their iron. Mycobacteria produce two types of siderophore, exochelins that are secreted and mycobactins that remain in the membrane. Mycobactins are complex molecules with aromatic and nitrogen-containing rings substituted with fatty acids. Thus they are lipid-soluble and reside within the membrane.

Iron availability is particularly important for pathogenic bacteria, which have to compete with host iron-binding proteins such as lactoferrin and transferrin. The ability to sequester iron plays a crucial role in the establishment of infection by pathogens, and the production of siderophores mirrors the virulence of different strains of bacteria such as *Salmonella typhimurium*.

6D   Conclusion

In this chapter we have described some of the important functional aspects of lipids. There are, however, many other areas where lipid composition seems to be specially adapted to help with life under specialised circumstances. Obvious examples of this are the unusual lipids and lipid compositions of salt-tolerant plants and bacteria and of the thermophilic micro-organisms of volcanic hot springs. To date, though, we do not know the exact molecular function of such fascinating adaptations. Perhaps the answers to these and other questions of lipid biochemistry will be solved by the readers of this book.

Further reading

Chapter 1

Rogers, H. J., H. R. Perkins and J. B. Ward 1980. *Microbial cell walls and membranes*. London: Chapman and Hall.

Woese, C. R., L. J. Magrum and G. E. Fox 1978. Archaebacteria. *J. Molec. Evolution* **11**, 245–52.

Chapter 2

Goldfine, H. 1982. Lipids of prokaryotes – structure and distribution. *Curr. Top. Memb. Transport* **17**, 1–43.

Hildritch, T. P. and P. N. Williams 1964. *The chemical constitution of natural fats*. London: Chapman and Hall.

Hitchcock, C. and B. W. Nichols 1971. *Plant lipid biochemistry*. London: Academic Press.

Kates, M. 1978. The phytanyl ether-linked polar lipids and isoprenoid neutral lipids of extremely halophilic bacteria. *Prog. Chem. Fats other Lipids* **15**, 301–42.

Nes, W. R. and W. D. Nes 1980. *Lipids in evolution*. New York: Plenum.

Nichols, B. W. 1973. Lipid composition and metabolism. In *The biology of blue-green algae*, 1st edn, N. G. Carr & B. A. Whitton (eds), 144–61. Oxford: Blackwell Scientific.

Ratledge, C and J. Stanford 1982. *The biology of the mycobacteria*, Vol. 1. New York and London: Academic Press. [Includes chapters on the composition, biosynthesis and functions of mycobacterial lipids.]

Stumpf, P. K. and E. E. Conn 1980. *Biochemistry of plants*, Vol. 4, *Lipids*. New York: Academic Press. [Various chapters on the composition, biosynthesis and function of plant lipids.]

Chapter 3

Eichenberger, W. 1977. Steryl glycosides and acylated steryl glycosides. In *Lipids and lipid polymers in higher plants*, M. Tevini & H. K. Lichtenthaler (eds) 169–81. Berlin: Springer.

Galliard, T. 1973. Phospholipid metabolism in photosynthetic plants. In *Form and function of phospholipids*, G. B. Ansell, R. M. C. Dawson & J. N. Hawthorne (eds), 253–88. Amsterdam: Elsevier.

Goodwin, T. W. 1980. *The biochemistry of the carotenoids*, 2nd edn, Vol. 1, *Plants*. London: Chapman and Hall.

Greenawalt, J. T. and L. Whiteside 1975. Mesosomes: membranous bacterial organelles. *Bact. Revs* **39**, 405–63.

op den Kamp, J. A. F. 1981. The asymmetric architecture of membranes. In *Membrane structure* J. B. Finean & R. H. Michell (eds), 83–126. Amsterdam: Elsevier.

Kolattukudy, P. E. 1976. *Chemistry and biochemistry of natural waxes*. New York: Elsevier.

Kushner, D. J. 1978. *Microbial life in extreme environments*. London: Academic Press.

Lechevalier, M. P. 1977. Lipids in bacterial taxonomy – a taxonomist's view. *CRC Crit. Rev. Microbiol.* **5**, 109–210.

Mudd, J. B. and R. E. Garcia 1975. Biosynthesis of glycolipids. In *Recent advances in the chemistry and biochemistry of plant lipids*, T. Galliard & E. I. Mercer (eds), 162–201. London: Academic Press.

Olson, J. M. 1980. Chlorophyll organisation in green photosynthetic bacteria. *Biochim. Biophys. Acta* **594**, 33–51.

Remsen, C. C. 1982. Structural attributes of membranous organelles in bacteria. *Int. Rev. Cytol.* **76**, 195–223.

Shaw, N. 1975. Bacterial glycolipids and glycophospholipids. *Adv. Microb. Physiol.* **12**, 141–67.

Spurgeon, S. L. and J. W. Porter 1980. Carotenoids. In *Biochemistry of plants*, Vol. 4, P. K. Stumpf & E. E. Conn (eds), 420–84. New York: Academic Press.

Thiele, O. W. and J. Oulerey 1981. Occurrence of phosphatidylcholine in hydrogen-oxidising bacteria. *Eur. J. Biochem.* **118**, 183–6.

Weete, J. D. 1980. *Lipid biochemistry of fungi and other organisms*. New York: Plenum.

Wintermans, J. F. G. M. and P. J. C. Kuiper 1982. *Biochemistry and metabolism of plant lipids*. Amsterdam: Elsevier.

Chapter 4

Bloch, K. and D. Vance 1977. Control mechanisms in the synthesis of saturated fatty acids. *A. Rev. Biochem.* **46**, 263–98.

Britton, G. 1982. Carotenoid biosynthesis in higher plants. *Physiol. Veg.* **20**, 735–55.

Clark, D. and J. E. Cronan, Jr 1981. Bacterial mutants for the study of lipid metabolism. *Method. Enzymol.* **72**, 293–307.

Cronan, J. E., Jr 1978. Molecular biology of bacterial membrane lipids. *A. Rev. Biochem.* **47**, 163–89.

Douce, R. and J. Joyard 1979. Structure and function of the plastid envelope. In *Advances in botanical research*, Vol. 7, H. Woolhouse (ed.), 1–116. New York: Academic Press.

Fischer, W. 1981. Glycerophosphoglycolipids, presumptive biosynthetic precursors of lipoteichoic acids. In *Chemistry and biological activities of bacterial surface amphiphiles*, G. D. Shockman & A. J. Wicken (eds), 209–28. New York: Academic Press.

Fulco, A. J. 1983. Fatty acid metabolism in bacteria. *Prog. Lipid Res.* **22**, 133–60.

Galanos, C., O. Lüderitz, E. T. Rietschel and O. Westphal 1977. Newer aspects of the chemistry and biology of bacterial lipopolysaccharides with special reference to their lipid A component. In *International review of biochemistry, Series II, Vol. 14, Biochemistry of lipids*, T. W. Goodwin (ed.), 239–335. Baltimore, Md: University Park Press.

Goodwin, T. W. 1980. Biosynthesis of sterols. In *Biochemistry of plants*, Vol. 4, P. K. Stumpf & E. E. Conn (eds), 485–507. New York: Academic Press.

Harwood, J. L. 1979. Synthesis of acyl lipids in plant tissues. *Prog. Lipid Res.* **18**, 55–86.

Harwood, J. L. 1980. Sulpholipids. In *Biochemistry of plants*, Vol. 4, P. K. Stumpf & E. E. Conn (eds), 301–20. New York: Academic Press.

Kawahara, K., K. Uchida and K. Aida 1979. Direct hydroxylation in the biosynthesis of hydroxy fatty acid in lipid A of *Pseudomonas ovalis*. *Biochim. Biophys. Acta* **572**, 1–8.

Kolattukudy, P. E. 1980. Cutin, suberin and waxes. In *Biochemistry of plants*, Vol. 4, P. K. Stumpf & E. E. Conn (eds), 571–645. New York: Academic Press.

Lekakis, N. J. 1977. α-Oxidation of fatty acids in *Escherichia coli*. *FEMS Microbiol. Lett.* **1**, 289–92.

Loomis, N. D. and R. Croteau 1980. Biochemistry of terpenoids. In *Biochemistry of plants*, Vol. 4, P. K. Stumpf & E. E. Conn (eds), 363–418. New York: Academic Press.

Mangold, H. K. and F. Spener 1980. Biosynthesis of cyclic fatty acids. In *Biochemistry of plants*, Vol. 4, P. K. Stumpf & E. E. Conn (eds), 647–63. New York: Academic Press.

Moore, T. S., Jr 1982. Phospholipid biosynthesis. *A. Rev. Plant Physiol.* **33**, 235–59.

Mudd, J. B. 1980. Phospholipid biosynthesis. In *Biochemistry of plants*, Vol. 4, P. K. Stumpf & E. E. Conn (eds), 249–82. New York: Academic Press.

Poralla, K. 1982. Considerations on the evolution of steroids as membrane components. *FEMS Microbiol. Lett.* **13**, 131–5.

Raetz, C. R. H. 1978. Enzymology, genetics and regulation of membrane phospholipid synthesis in *Escherichia coli*. *Microbiol. Revs.* **42**, 614–59.

Rock, C. O. and J. E. Cronan, Jr. 1982. Regulation of bacterial membrane lipid synthesis. *Curr. Top. Memb. Transport* **17**, 207–33.

Roughan, P. G. and C. R. Slack 1982. Cellular organisation of glycolipid metabolism. *A. Rev. Plant Physiol.* **33**, 97–132.

Schulman, H. and E. P. Kennedy 1977. Regulation of turnover of membrane phospholipids in synthesis of membrane-derived oligosaccharides of *Escherichia coli*. *J. Biol. Chem.* **252**, 4250–5.

Snyder, F. 1972. *Ether lipids: chemistry and biology*. New York: Academic Press.

Stumpf, P. K. 1980. Biosynthesis of saturated and unsaturated fatty acids. In *Biochemistry of plants*, Vol. 4, P. K. Stumpf & E. E. Conn (eds), 177–204. New York: Academic Press.

Stumpf, P. K., T. Shimakata, K. Eastwell, D. J. Murphy, B. Liedvogel, J. B. Ohlrogge and D. N. Kuhn 1982. Biosynthesis of fatty acids in a leaf cell. In *Biochemistry and metabolism of plant lipids* J. F. G. M. Wintermans & P. J. C. Kuiper (eds), 3–12. Amsterdam: Elsevier.

Thompson, G. A. (1980. *The regulation of membrane lipid metabolism*. Boca Raton, Fla: CRC Press.

Wakil, S. J., J. K. Stoops and V. C. Joshi 1983. Fatty acid synthesis and its regulation. *A. Rev. Biochem.* **52**, 537–79.

Chapter 5

Finnerty, W. R. 1978. Physiology and biochemistry of bacterial phospholipid metabolism. *Adv. Microb. Physiol.* **18**, 177–233.

Galliard, T. 1980. Degradation of acyl lipids: hydrolytic and oxidative enzymes. In *Biochemistry of plants*, Vol. 4, P. K. Stumpf & E. E. Conn (eds), 85–116. New York: Academic Press.

Galliard, T. and H. W-S. Chan 1980. Lipoxygenases. In *Biochemistry of plants*, Vol. 4, P. K. Stumpf & E. E. Conn (eds), 131–61. New York: Academic Press.

Hobson, P. N. and R. J. Wallace 1982. Microbial ecology and activities in the rumen: Part II. *CRC Crit. Rev. Microbiol.* **9**, 253–320.

Simons, R. W., K. T. Hughes and W. D. Nunn 1980. Regulation of fatty acid degradation in *Escherichia coli*: dominance studies with strains merodiploid in gene *fad* R. *J. Bact.* **143**, 726–30.

Chapter 6

Appelqvist, L-A. 1975. Biochemical and structural aspects of storage and membrane lipids in developing oil seeds. In *Recent advances in the chemistry and biochemistry of plant lipids*, T. Galliard & E. I. Mercer (eds), 247–85. London: Academic Press.

Beevers, H. 1980. The role of the glyoxylate cycle. In *Biochemistry of plants*, Vol. 4, P. K. Stumpf & E. E. Conn (eds), 117–30. New York: Academic Press.

Cullis, P. R. and B. de Kruijff 1979. Lipid polymorphism and the functional role of lipids in biological membranes. *Biochim. Biophys. Acta* **559**, 399–420.

Finean, J. B. and R. H. Michell 1981. *New comprehensive biochemistry*, Vol. 1, *Membrane structure*. Amsterdam: Elsevier.

Gurr, M. I. 1980. The biosynthesis of triacylglycerols. In *Biochemistry of plants*, Vol. 4, P. K. Stumpf & E. E. Conn, 205–49. New York: Academic Press.

Hatch, M. D. and N. K. Boardman (volume eds) 1981. *Biochemistry of plants*, P. K. Stumpf and E. E. Conn (eds), Vol. 8, *Photosynthesis*. New York: Academic Press.

Kates, H. and A. Kuksis 1980. *Membrane fluidity. Biophysical techniques and cellular recognition*. Clifton, N.J.: Humana Press.

Lederer, E. 1976. Cord factor and related trehalose esters. *Chem. Phys. Lipids* **16**, 91–106.

McElhaney, R. N. 1976. The biological significance of alterations in the fatty acid composition of microbial membrane lipids in response to changes in environmental temperature. In *Extreme environments. Mechanisms of microbial adaptation*, M. R. Heinrich (ed.), 255–81. New York: Academic Press.

McElhaney, R. N. 1982. Effects of membrane lipids on transport and enzyme activities. *Curr. Top. Memb. Transport* **17**, 317–80.

Overath, P. and L. Thilo 1978. Structural and functional aspects of biological membranes revealed by lipid phase transitions. In *International review of biochemistry*, Series II, Vol. 19, *Biochemistry of cell walls and membranes*, J. C. Metcalfe (ed.), 1–44. Baltimore, Md: University Park Press.

Quinn, P. J. 1981. The fluidity of cell membranes and its regulation. *Prog. Biophys. Molec. Biol.* **38**, 1–104.

Quinn, P. J. and W. P. Williams 1983. The structural role of lipids in photosynthetic membranes. *Biochim. Biophys. Acta* **737**, 223–6.

Raison, J. K. 1980. Membrane lipids: structure and function. In *Biochemistry of plants*, Vol. 4, P. K. Stumpf & E. E. Conn (eds), 57–83. New York: Academic Press.

Rothfield, L. and D. Romeo 1971. Role of lipids in the biosynthesis of the bacterial cell envelope. *Bact. Revs*. **35**, 14–38.

Russell, N. J. and J. L. Harwood 1979. Changes in the acyl lipid composition of photosynthetic bacteria grown under photosynthetic and non-photosynthetic conditions. *Biochem. J.* **181**, 339–45.

Sanderman, H. 1978. Regulation of membrane enzymes by lipids. *Biochim. Biophys. Acta* **515**, 209–37.

Index

References to text sections are given in **bold** type, and text figure numbers are given in *italics*.